T0350135

ELEMENTS OF NUMERICAL ANALYSIS WITH MATHEMATICA®

ELEMENTS OF NUMERICAL ANALYSIS WITH MATHEMATICA®

John Loustau

Hunter College, City University of New York, USA

NEW JERSEY • LONDON • SINGAPORE • BEIJING • SHANGHAI • HONG KONG • TAIPEI • CHENNAI • TOKYO

Published by

World Scientific Publishing Co. Pte. Ltd.
5 Toh Tuck Link, Singapore 596224
USA office: 27 Warren Street, Suite 401-402, Hackensack, NJ 07601
UK office: 57 Shelton Street, Covent Garden, London WC2H 9HE

Library of Congress Cataloging-in-Publication Data
Names: Loustau, John, 1943– author.
Title: Elements of numerical analysis with Mathematica / by John Loustau
(Hunter College, City University of New York, USA).
Description: New Jersey : World Scientific, 2017. | Includes bibliographical references and index.
Identifiers: LCCN 2017020923| ISBN 9789813224155 (hardcover : alk. paper) |
ISBN 9789813222717 (pbk : alk. paper)
Subjects: LCSH: Numerical analysis--Study and teaching (Higher) |
Numerical analysis--Data processing. | Mathematica (Computer file)
Classification: LCC QA297 .L68 2017 | DDC 518--dc23
LC record available at https://lccn.loc.gov/2017020923

British Library Cataloguing-in-Publication Data
A catalogue record for this book is available from the British Library.

Printed in Singapore

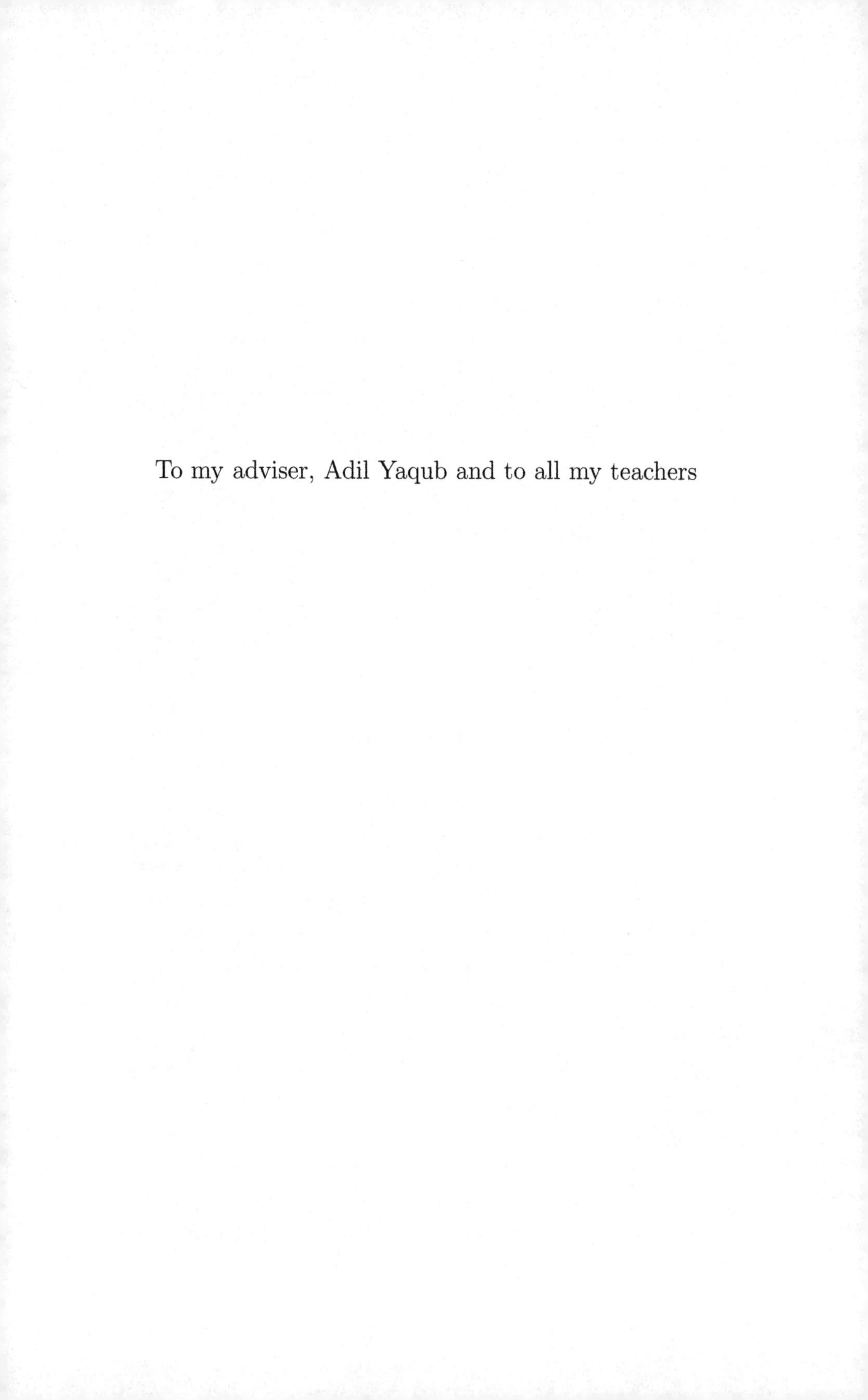

To my adviser, Adil Yaqub and to all my teachers

Preface

This book is an introduction to numerical analysis. It is intended for third or fourth year undergraduates or beginning master level students. The required background includes multivariate calculus and linear algebra. Some knowledge of real analysis is recommended. In terms of programming, some programming experience would be a plus but is not required. Indeed, the book is self-contained in this respect. But, the treatment is likely too fast for some beginners.

The programming environment is *Mathematica*, a 4GL with an advanced symbolic manipulation component. We make an effort to keep the programming component simple. In particular, we use as little of the programming language as possible. We are currently using Version 10. However, this material has been developed and used over a period of years. There are some programming changes or additions included in more recent versions. But none of these affect the programs included in this text. With Version 10, *Mathematica* is also referred to as *Wolfram Language*.

The purpose is to introduce numerical analysis. Since the post WWII period, [Grcar (2011)], numerical analysis has been the mathematics supporting large calculations carried out on a computer. Hence, a course in this topic must include computational exercises and these should be large enough or sufficiently complex to warrant the use of a computer. It is better still if there are realistic problems; ones that the student can imagine would arise in an actual application.

One dominant application of numerical techniques concerns simulating processes represented by differential equations. In this setting, we are charged with estimating the existent but unknown solution to a differential equation. We have organized the topics of this book to introduce some of the classical approaches to this problem. In particular, we develop finite difference method (FDM) for a parabolic equation in one spatial variable, including explicit, implicit and Crank-Nicolson FDM. In addition, we touch on stability. In another direction, we present some of the elementary techniques used to simulate the solution of an ordinary differential equation. Monte Carlo method is another means to simulate a differential equation. We present the basics of this method. We include the simulation of a stochastic differential equation.

Another area of current interest is certain *Big Data* applications. The solution procedures for problems that arise in this area include solving large linear systems of equations. Generally, these systems are too large for Gauss-Jordan elimination. We will develop two basic solution procedures including Gauss-Seidel. In this context, we introduce the student to the basics of Krylov subspaces. Another aspect of Big Data applications is optimization, multivariate max/min problems. This is also an important part of our development. We include both the *greatest descent* and *Hessian* variants.

We present the techniques of numerical analysis together with the supporting theory. Much of this book is organized in traditional definition, theorem, proof format. With a small number of exceptions, the theory is self-contained within this text and prerequisites. In the exceptional cases, the necessary supporting material is identified and referenced. But, full understanding of the material requires both knowledge of the mathematical foundations and hands on experience. In this regard, we include examples and exercises that are beyond what can be done with pencil and paper. In this way, we emphasize the natural link between numerical analysis and computing.

We present the material of the text in the given order. It is intended to be a one semester course. Programming in *Mathematica* is sufficiently intuitive for students with math or engineering background that little or no programming background is necessary. In fact, the reader's experience manually solving problems provides the necessary foundation toward programming sophisticated processes. At the same time, we find it useful to schedule computer lab time for the first few weeks of the semester. In this facility, we can provide one-on-one programming support. By mid-semester all students are on an even footing programming-wise.

For most students this class is a rewarding experience. In this setting, they are able to solve problems that are realistic in scale and complexity. To this point, their prior experience is often restricted to problems that can be executed easily with pencil and paper.

In many respects a course based on this text would function as a capstone course for the undergraduate. This is because it calls upon so much of a students' undergraduate experience, calculus, single and multivariate, vector calculus, linear algebra, real analysis, ODE and PDE and probability.

This text arose from our need of a beginning numerical analysis text that was sufficiently mathematical for our students and supported *Mathematica* as the programming platform. During that time, our students have seen topics come and go as we settled on a stable course. Without their participation this text could not have been written. Special acknowledgement goes to those who helped me understand how to present this material. In particular, this includes Scott Irwin, Yevgeniy Milman, Andrew Hofstrand, Evan Curcio, Gregory Javens and James Kluz.

Special acknowledgment goes to those who helped me understand how to present this material. In particular, this includes Scott Irwin, Yevgeniy Milman, Andrew Hofstrand, Evan Curcio, Gregory Javens, Hassan Mahmood and James Kluz.

About the Cover: Carl Friedrich Gauss was a giant among mathematicians. There are several signature procedures attributed to Gauss that arise in the text. For instance, Gaussian quadrature is often considered the *gold standard* for numerical integration. In general, problems stated here are resolved as systems of linear equations. Moreover, the basic techniques we use today are attributed to Gauss. Hence, we may think of Gauss as the central figure in the development of this topic.

John Loustau
Hunter College (CUNY)
New York, 2017

Contents

Chapter 1

Beginnings

Introduction

This chapter provides a brief introduction to programming for those who have never programmed. For those with programming experience, this is the introduction to *Mathematica.* One of the several advantages to using a 4GL such as *Mathematica* is that it makes numerical methods accessible to all students with multivariate calculus and linear algebra. Indeed, most students catch on very quickly to programming in *Mathematica* and are doing complicated programs well before the end of the semester course. To support the learning process, there are tutorials available by selecting *Help* from the system menu and then *Documentation Center.* The first item available in documentation contains the basic tutorials. For those who prefer hard copy references, there are several textbooks available from booksellers.

A second feature of this chapter is to introduce the reader to the quirks of *Mathematica. Mathematica* was originally developed by mathematicians, and therefore it has a mathematician's point of view. If your background is with C++ or another of the computer scientist developed programming products, you will find *Mathematica* to be similar on the surface but significantly different at lower levels. If you have never programmed in an interpreted 4GL, then you have something to get used to.

In this chapter, we introduce the terminology associated to computer error. Computers must represent decimal values in a finite number of significant digits. Therefore, the representation is often only close to the actual value. For instance $1/3$ is not 0.3333333. The error inherent in the representation can be magnified during normal arithmetic operations. In extreme cases, this may yield ridiculous results. When using any computer system you must always be cognizant of the potential of error in your calculations. We will see an example of this in Section 1.2.

We next look at Newton's method. Most calculus courses include Newton's method for finding roots of differentiable functions. If you have solved a Newton's method problem with pencil and paper, you know that doing two or three iterations of the process is a nightmare. Even the simplest cases are not the sort of thing most students want to do. Now, we see that it is easy to program. In this regard, it is an excellent problem for the beginning student. In addition, *Mathematica* provides a built in function that performs Newton's method. It is empowering for the student to compare his results to the output produced by *Mathematica.* We follow Newton's method by the secant method to find the root of a function. This provides the student with the first example of an error estimating procedure.

By the end of the chapter, the student should be able to program the basic arithmetic operations, access the standard mathematical functions, program loops and execute conditional statements (if ... then ... else ...). A special feature of *Mathematica* is the graphics engine. With minimal effort, the student can display sophisticated graphical output. By the end of this chapter the student will be able to use the basic 2D graphics commands.

1.1 The Programming Basics for *Mathematica*

We use *Mathematica* Version 10. Each year when the university renews its license, the version changes. In the past, programs for one version are either fully upgradeable to the subsequent version or Wolfram provides a program that upgrades program code written for one version to the next.

During this semester you will be programming in *Mathematica.* To begin with, you will learn to be able to program the following.

(1) The basic arithmetic operations (addition, subtraction, multiplication, division, exponentiation and roots)
(2) Define a function
(3) Loops (Do Loop or While Loop)
(4) Conditionals (If ... then ... else)
(5) Basic graphics (point plot, function plot, parametric plot)

We begin with item 1. You add two numbers $3 + 5$, subtract $3 - 5$, multiply $3 * 5$ and divide $3/5$. Alternatively, if $a = 3$ and $b = 5$, then $a + b$, $a - b$, $a * b$ and a/b have exactly the same result. For exponents, $a^b = 243$ and $b^a = 125$. The usual math functions cosine, Cos[x], sine,

Sin[x], exponential, Exp[x] and so forth all begin with a capital letter. The argument is enclosed in square brackets.

To start, bring up Mathematica and select 'new notebook'. Now type a line or two of program code. For instance any two of the statements in the prior paragraph. To execute the code, you hold the *shift* key down and then press *enter*.

Consider the polynomial in 2 variables, $f(x, y) = (x + y)^2 - 2xy - y^2$. You define this in a *Mathematica* program with the following statement,

```
f[x_,y_] = (x+y)^2 - 2*x*y - y^2
```

You can find descriptions of each of these in the first two or three chapters of most any programming text for *Mathematica*. In addition *Mathematica* includes a programming tutorial accessible via the *Help Menu*.

There are some comments that need to be made.

A. Error messages in *Mathematica* are cryptic at best. After a while you will begin to understand what they mean and use them to debug your program. But this will take some experience. On the other hand there are circumstances where you might expect to receive an error or warning message but none is generated. For instance,

```
If [x == 0,
    x = 5
    ];
```

will test the value of x. If it is zero, then it will be set to 5. On the other hand,

```
If [x = 0,
    x = 5
    ];
```

sets x to zero and then to 5, no matter what value x may have.

B. When multiplying two variables denoted by a and b, you may write $a * b$ or $a\ b$ (with a space between). But if you write ab (without a space), then *Mathematica* will think that ab is a new variable and not the product of two variables. Because of how *Mathematica* spaces input, it is not always easy to distinguish between $a\ b$ and ab. You will save yourself a lot of time if you always use the asterisk for multiplication.

C. When you write mathematics, you may use the three grouping symbols, parentheses ( ), square brackets [ ], and braces {}, interchangeably. In *Mathematica* you may not do this. Parenthesis may only be used in computations for grouping. Square brackets are only used around the independent variables of a function, while braces are only used for vectors and matrices. For instance each of the following expressions will cause an error in a *Mathematica* program:

$$[a+b]^2, f(x), (x,y).$$

The correct expressions are

$$(a+b)^2, f[x], \{x,y\}.$$

D. When defining a function in your program, you always follow the independent variable(s) with an underscore. This is how *Mathematica* identifies the independent variable. Later when you reference the function, you must not use the underscore. For instance the following statements defines a function as xe^{-x}, evaluates the function at $x = 1$ and then defines a second function as the derivative of the first.

$$f[x\_] = x * Exp[-x];$$

$$y = f[1];$$

$$g[x\_] = D[f[x], x];$$

Once you have used a letter or group of letters for the independent variable in a function definition, it is best to not use it for any other purpose. For instance, we have used x as the independent variable for both f and g. We can now write $f[4]$ to evaluate f at 4 and we can write $a = 4$, $f[a]$. But if we write $x = 4$ then we will have a clash. On the one hand x is the independent variable for f, and on the other hand, it is a constant 4.

E. All reserved words in *Mathematica* begin with upper case letters. When you define a function or a variable, it is best to use names that begin with lower case letters. This way functions and variables you define will not clash with the ones *Mathematica* has reserved. For instance, Pi is 3.14...., and I is $Sqrt[-1]$. Any attempt to use these symbols for any other purpose will at the very least cause your output to be strange.

F. There are two ways to indicate the end of a line of program code. First, you may simply hit the 'return' key and go to the next line. Second, you may end the line with a semicolon. The result upon execution is slightly different. In the first case the result of the line of code is printed when the code is executed. In the second the output of the line of code is suppressed. For instance, if you enter

$$a = 3 + 5$$

and execute (press shift+enter), *Mathematica* will return the value, 8. On the other hand if you type

$$a = 3 + 5;$$

and execute, then there is no printed output. In any event, the calculation does occur and the result is stored in a. Any subsequent calculation can access the result by referencing a.

G. It is best to have only one line of program code per physical line on the page. For short programs, violating this rule should not cause any problems. For long and involved programs, debugging is often a serious problem. If you have several lines of code on the same physical line, you may have trouble noticing a particular line of code that is causing an error. For instance,

$$z = x + y;$$
$$x = z + 1;$$

is preferred,

$$z = x + y; \quad x = z + 1;$$

is not.

H. *Mathematica* has distinct computational procedures for integer arithmetic and decimal arithmetic. Integer calculations take much longer to execute. The result, expressed as fractions, is exact. Decimal calculations are much faster but there is round off error. (See Section 2.) For instance, if all the data for a program is whole numbers and is entered without decimal point, then *Mathematical* will assume that the calculations are integer and proceed. (See problems 1 and 2 below.)

I. We did not use := when defining the function $f(x, y)$. There are technical differences between the two symbols used to define a function. Different authors will suggest one or the other. Our take on this is that := is used when defining a module, a function defined as a sub-program and accessed at several different locations in your program. Otherwise, to define a simple function as we have done, it is best to use =. A problem arises if you need to use a functions name for several different functions in the same program. For instance, f may refer to a polynomial with coefficients that change during the execution of the program. If you use := to define f, then you will need to use *Clear* before you can change the coefficients. If you use = to define f, then the problem will not arise.

Exercises:

1. Write a program that defines the function $f(x) = x^{-2} + x^2$ and evaluates f at $x = 2$ and $x = 3$. Use the *Mathematica* function *Print* to display the output.

2. Repeat Exercise 1 for $x = 2.0$ and $x = 3.0$.

3. Define $g = f'$, the derivative of f. Evaluate g at 5 and 5.0.

1.2 Errors in Computation

Errors arise from several sources. There are the errors in data collection and data tabulation. This sort of data error needs to be avoided as much as possible. Usually this is accomplished by quality assurance procedures implemented at the team management level. This is not our concern in numerical methods. Programming errors are a quality control issue that is addressed in software engineering. These errors are avoided by following good practices of software engineering. For our own programs, we are best advised to be as simple as possible: simple in program design, simple in coding. A mundane program is much easier to control and modify than a brilliant but somewhat opaque one. The simple one may take longer to code or longer to execute, but it is usually preferred.

There are errors that arise because of the processes we use and the equipment that we execute on. Both are errors due to the discrete nature of the digital computer. These errors cannot be prevented and hence must be controlled via error estimation.

First, the computer cannot hold an infinite decimal. Hence, the decimal representation of fractions such as $1/3$ and $2/3$ are inherently incorrect. Further, subsequent computations using these numbers are incorrect. A small error in the decimal representation of a number when carried forward through an iterated process may accumulate and result in an error of considerable size. For instance, when solving a large linear system of equations, an error introduced in the upper left corner will iterate through the Gauss-Jordan process causing a significant error in the lower right corner. Hence, the larger the linear system, the larger the accumulated error.

Another type of error arises from discrete processes. For instance, suppose you have an unknown function $f(x)$. Suppose also that you know that $f(1) = 1$ and $f(1.1) = 1.21$. Then it is reasonable to estimate the derivative by the Newton quotient

$$\frac{df}{dx}(1) \approx \frac{1.21 - 1}{1.1 - 1} = \frac{0.21}{0.1} = 2.1.$$

If in fact, $f(x) = x^2$, then our estimated derivative is off by 0.1. But without knowing the actual function, we have no choice but to use the estimation. We are faced with one of two alternatives, doing nothing or proceeding with values that we expect are flawed. The only reasonable alternative is to proceed with errors provided we can estimate the error.

This is a special case of a more general problem. Suppose we want to compute a value y, but in fact, we can only compute the values at a sequence x_n that converges to x. Since we can never do our computation all the way to the limit, then we must have a means to estimate the nth, $x - x_n$.

We formalize the error in the following definition.

Definition 1.2.1. Suppose that there is a computation and that estimates a value x with the computed value $\tilde{x}$, then $e = x - \tilde{x}$ is called the *error*. In turn, $|x - \tilde{x}|$ is called the *absolute error*. If $x \neq 0$, then $(x - \tilde{x})/x$ is called the *relative error*. In this case, the *relative absolute error* is given by $|x - \tilde{x}|/|x|$.

It is reasonable to ask why we should care about e if we already know x. Indeed, if we know x, there is no need for a numerical technique to estimate x with $\tilde{x}$. In numerical analysis, the basic assumption is that x is not computable but estimable. Therefore, it is useful to have precise definitions for these terms, as there are situations where we can estimate the error e without knowing the actual value x.

Each numerical process should include a procedure to estimate the error. It can be argued that any procedure that does not include an error estimate is of no value. What is the purpose of executing a computation, if we have no idea whether or not the computed data approximates the actual value?

A second comment is in order. It is preferable to use the relative or relative absolute error. This is because these values are dimensionless. For instance, consider the example of the derivative of the squaring function. If the data is given in meters, then $e = 0.1$ *meters*. If the data were instead displayed in kilometers then $e = 0.001$ and or centimeters, $e = 10$. Even though the error is the same, the impression is different. However, for relative error the value is 0.1, is independent of the unit of measurement. When this is the case, we say that the data is *dimensionless*.

In Exercise 1 below, you are asked to execute a simple calculation which should always yield the same result independent of the input data. In this problem you are asked to use several different values of x. Unexpectedly, the results will vary broadly. In a simple calculation like this it is possible to look at how the numbers are represented and determine why the error occurs. But in any actual situation the calculations are so complex that such an analysis is virtually impossible. When executing a computation, you should always have an idea of what the results should be. If you get impossible output and there is no error in your program, then you may be looking at small errors in numerical representation compounded over perhaps thousands of separate arithmetic operations.

Exercises:

1. Consider the function $f(x,y) = ((x+y)^2 - 2xy - y^2)/x^2$. We expect that if $x \neq 0$, then $f(x,y) = 1$. Set $y = 10^3$ and compute f for $x = 10.0^{-1}$, 10.0^{-2}, 10.0^{-3}, 10.0^{-4}, 10.0^{-5}, 10.0^{-6}, 10.0^{-7}, 10.0^{-8}. For each value of x compute the absolute error.

2. Repeat Problem 1 with $g(x,y) = (x+y)^2/x^2 - 2xy/x^2 - y^2/x^2$. Why are the results different? Is $g[x, 10^3]$ the same function as $f[x, 10^3]$?

3. Repeat Problem 1 using $x = 10^{-1}$, 10^{-2}, 10^{-3}, $10^{(-4}$, 10^{-5}, 10^{-6}, 10^{-7}, 10^{-8}. Why are the results different?

1.3 Newton's Method

Suppose that you have a function $f(x) = y$ and want to find a root or a zero for f. Recall that $\tilde{x}$ is a root of f provided $f(\tilde{x}) = 0$. If f is continuous and $f(x_1) > 0$ and $f(x_2) < 0$ then you know that f must have at least one root between x_1 and x_2. This result, *The Intermediate Value Theorem*, is usually stated as early as Calculus 1 and most commonly proved in the first semester of Real Analysis. [Rudin (1976)]. There is an intuitively simple but inefficient means to determine a good approximation for $\tilde{x}$ based on this theorem.

(1) Consider the midpoint of the interval $[x_1, x_2]$, $(x_1 + x_2)/2 = \hat{x}$.
(2) If $f(\hat{x}) = 0$, then $\hat{x}$ is a root. Exit.
(3) If $f(\hat{x}) > 0$, then replace x_1 with $\hat{x}$.
(4) If $f(\hat{x}) < 0$, then replace x_2 with $\hat{x}$.

(5) Return to Step 1.

The following *Mathematica* code segment demonstrates this process. We use the fact that $a * b > 0$ if and only if a and b have the same sign.

```
test-root = x1;
test-value = 10^-5;
While[Abs[f[test-root]] > test-value,
   test-root = (x1 + x2)/2;
   If[f[test-root]*f[x1] >= 0,
      x1 = test-root,
      x2 = test-root
      ];
 ];
Print[test-root];
```

In this code fragment, we assume that f, $x1$ and $x2$ are already known. Notice that we use 10^{-5} as the *kick out threshold*. In particular, as soon as the absolute value of f at the current test-root is less than 10^{-5}, the process stops and the approximate root is printed. Since the procedure may only approximate the solution, then we do need a kick out threshold. The value $f(\hat{x})$ is called the *residual*, and we say that we are using a *residual kick out test*. Alternatively, if x_n and x_{n+1} are two successive computed values, then $|x_n - x_{n+1}| < 10^{-5}$ may be used to end the computation. The idea behind this kick out test is that the sequence of computed approximate roots is convergent and hence Cauchy. In particular, $|x_n - x_{n+1}| \to 0$.

There are two statements in the *Mathematica* program that are new to us. First there is an *If Then Else* statement. This is called a *conditional statement*. The statement begins with a condition. Following the condition is a block of program statements to be executed if the condition holds followed by a comma and then a block of program statements to be executed if the condition fails. In this case the condition is that the sign of f at x_1 and the current approximate root is the same. If the condition is satisfied, then replace x_1 by the current value. If it is false, then replace x_2 by the current value.

There is also a *While Loop*. Following the *While* there is a condition. The *While loop* will execute as long as the condition holds. When it fails, the program will exit the loop. In this case the condition compares the absolute residual to the kick out threshold.

A more efficient means to locate an approximate root is called *Newton's method*. You probably saw this in Calculus. In this case, we suppose that f is differentiable and that we want to find a root for f near x_1. The idea is that the tangent line for f at x_1 is a good approximation for f near x_1. The equation for the tangent is a degree one polynomial. It is easily solved to locate a root. We call the root that we call x_2. If x_2 is closer to the root of f than x_1, then the method has been productive.

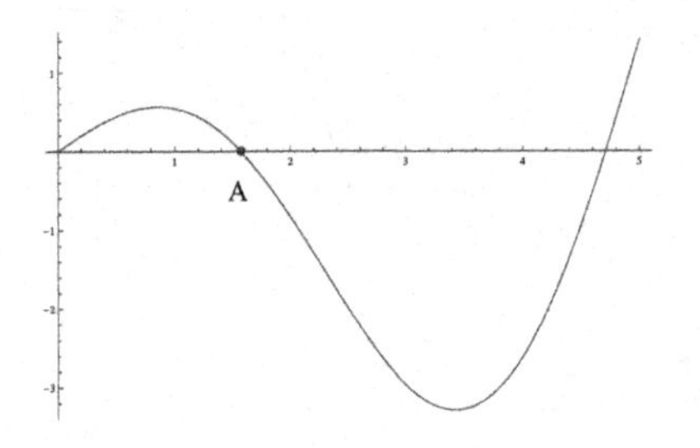

Figure 1.3.1: $f(x) = x\cos(x)$

Figure 1.3.2: f and the tangent line at $B = (2, f(2))$

For instance, if $f(x) = x\cos(x)$, then there is a root at the point A near $x = 1.6$. (See Figure 1.3.1.)

If we start the process with $x = 2$, $f'(2) \approx -2.2347$, $f(2) \approx -0.8233$ and the tangent to f at 2 is given by $h(x) = f'(2)(x-2) + f(2)$. Now h crosses the x-axis at 1.62757. Figure 1.3.2 shows f together with the tangent.

If we write this out formally, then starting at x_1, the tangent line has slope $f'(x_1)$ and passes through the point $(x_1, f(x_1))$. Hence,

$$f'(x_1) = \frac{y - f(x_1)}{x - x_1}.$$

Setting $y = 0$ and solving for x, we get

$$f'(x_1)(x - x_1) = -f(x_1),$$

or

$$x = x_1 - \frac{f(x_1)}{f'(x_1)}. \tag{1.3.1}$$

Replacing x_1 by x we get an iterative process that we can repeat until the absolute value of $|f(x_1)|$ is less than the kick out threshold value. We will call (1.3.1) the *operative statement.* The following steps provide an outline for the program.

(1) Provide an initial estimate for the root, x_1.
(2) Set up a *while loop*, use a residual threshold test to kick out.
(3) Put the operative statement inside the loop.
(4) Close the loop.
(5) Print the result.

The initital estimate for the root is often called the *seed.*

Newton's method is implemented in *Mathematica* via the *FindRoot* command. For instance, $f(x) = x\cos(x)$ has a root between $x = 1$ and 3. The following *Mathematica* statement will implement Newton's method to get an approximate value for the root.

```
x-root = FindRoot[x*Ccos[x] == 0, {x, 2}];
```

If you do a search on *FindRoot* in *Mathematica help*, you will see that there are options available to the programmer. Among the options is the opportunity to set the residual kick out threshold or use the default kick out value provided by *Mathematica.* In addition, you may set the maximal number of iterations for the process or accept the *Mathematica* preset iteration count limit. The idea is to stop the process after a set number of iterations no matter the value of the residual.

We can add an iteration count limit to our basic Newton's method program. To accomplish this, we will need to set two new variables, *iter-limit* and *iter-cnt.* Suppose we want to stop the process after 10000 iterations no matter what the value of the residual. We accomplish this with a compound conditional on the *While* statement. Then we need only add 5 new statements to our program. In *Mathematica* this would look like,

```
While[f[x]>10^{-5} && iter-cnt < iter-limit,
```

This reads, while the residual is greater than 10^{-5} and the iterations is less than 10000.

(1) Provide a seed, an initial estimate for the root, x_1.
(2) Set *iter-limit* = 10000.
(3) Set *iter-cnt* = 0
(4) Set up a *while loop* as outlined above. Use a residual threshold test to kick out plus an iteration count limit.
(5) Put the operative statement inside the loop.
(6) Add 1 to *iter-cnt*
(7) Close the loop.

(8) Print the result.
(9) Print *iter-cnt*

Before passing on, several words of caution are in order. First, if $f'(x_1) = 0$ then the process will fail. Indeed, in this case, the tangent line is parallel to the x-axis. Second, if $f'(x_1)$ is not zero, but nearly zero, then x given by (1.3.1) will be far from x_1. Indeed, for $f(x) = x\cos(x)$, $x_1 = 0.9$, then the slope of f at x_1 is approximately 0.0833 and the process will locate the root near 8. (See Figure 1.3.3.) Hence, it is important that $f'(x_1)$ be bounded away from zero.

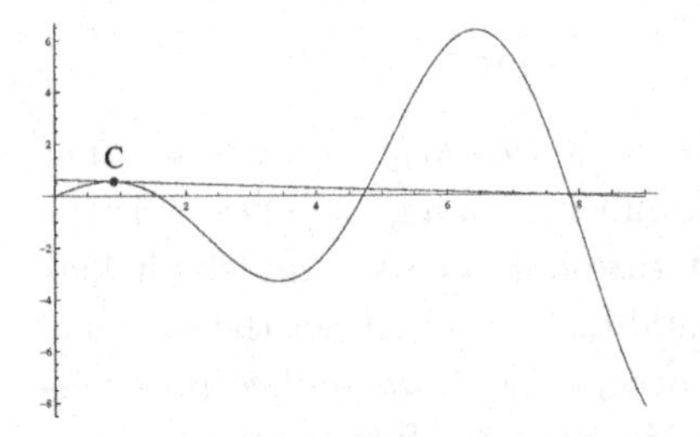

Figure 1.3.3: f and the tangent line at $C = (x_1, f(x_1))$

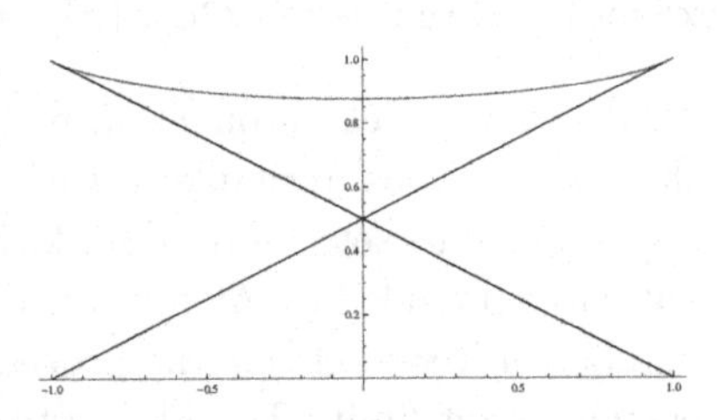

Figure 1.3.4: A Newton's method, cyclic order 2

Finally, it is possible that Newton's method will fail to find any approximate root. Indeed, the process may cycle. In particular, starting at a value x_1 you may pass on to a succession of values $x_2, x_3, ..., x_n$ only to have $x_n = x_1$. Once you are back to the original value, then the cycle is set and further processing is useless. Figure 1.3.4 shows a function f where $f(1) = f(-1) = 1$, $f'(1) = -(1/2)$ and $f'(-1) = 1/2$. Hence, $x_1 = 1$, $x_2 = -1$, $x_3 = 1$ and so forth. The example is constructed using a Bezier curve. It is indeed a function but not the sort of thing one would expect under normal circumstances. Later when we develop Bezier curves, we will revisit this example. It is now apparent why an iteration count kick out is essential.

Exercises:

1. Consider the function $f(x) = x^2$. We know that f has a root at zero. Suppose we select the seed to be 0.5 and the threshold to be 10^{-5}. Execute a program in *Mathematica* to estimate the root. Since you already

know the outcome, this sort of example is a good means to verify that your program is correct.

2. Plot the function $f(x) = xe^{-x} - 0.16064$.

a. Use *FindRoot* to locate a root near $x = 3$.

b. Write a program in *Mathematica* implementing Newton's method. Use your program to approximate the root near $x = 3$. Set the maximal number of iterations to 100 and the residual kick out threshold to 10^{-5}. How many iterations does your program actually execute before it stops?

c. Redo (b) with a kick out threshold set to 10^{-2}. Using the result of (a) as the actual and this result as the computed, calculate the relative absolute error.

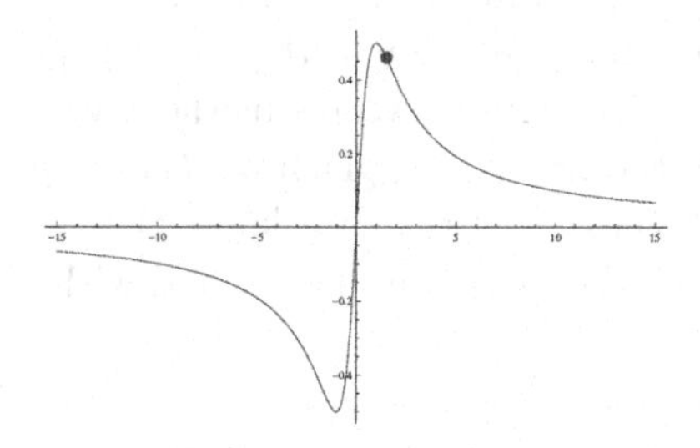

Figure 1.3.5: f and the tangent line at $C = (x_1, f(x_1))$

3. Figure 1.3.5 shows the graph of $f(x) = x/(x^2 + 1)$ together with the point $(1.5, f(1.5))$.

a. Use FindRoot to solve $f(x) = 0$ starting at 1.5. What happens? Why?

b. Write your own program to execute Newton's method starting at $x = 1.5$. What is the output for the first 10 iterations?

c. Plot f along with the tangent at the 4th iteration. Put both plots on the same axis.

Hint: Execute both plots separately and save the output in a variable. Then execute the *Show* statement. For instance, for tangent $h(x)$,

```
funplt = Plot[f[x], {x, -1, 10}];
tngplt = PlotOh[x], {x, -1, 10}];
Show[funplt, tngplt, PlotRange->All]
```

4. Consider the function $f(x) = \cos(e^x)$.

a. Define the function and plot it on the interval [-1, 4].

b. Locate a possible root, identify the seed and approximate the root via your implementation of Newton's method. Coimpare your estimate to *FindRoot*.

1.4 Secant Method

The secant method is a second technique to estimate the zero of a function. It is closely related to Newton's method. On the one hand, it is less efficient, while on the other, it is free of the anomalies which may arise with Newton's method. In particular, it will always locate an approximate root and the root will be in the intended vicinity. To describe the secant method, we continue the notation begun in the last section.

For the secant method we begin with a known positive and negative value for the function. As with Newton's method, we call these points the *seeds*. Suppose that the function is given by $f(x) = y$. If f is continuous and $f(x_1) > 0$ and $f(x_2) < 0$ then we know that f must have at least one root between x_1 and x_2. The secant method with seeds x_1 and x_2 will estimate that root.

The method is simple. It proceeds by considering the line connecting $(x_1, f(x_1))$ and $(x_2, f(x_2))$. We let $\hat{x}$ denote the point where the line (or secant) intersects the x-axis. If $f(\hat{x}) = 0$, then we have the root between x_1 and x_2. If $f(\hat{x}) > 0$, then we replace x_1 with $\hat{x}$ and proceed. Otherwise, we replace x_2 with $\hat{x}$.

For instance if $f(x) = x\cos(x)$, then there is a root between 1 and 2. Setting $x_1 = 1$ and $x_2 = 2$, then the line is given by $\lambda(x) = [(f(2) - f(1))/(2-1)](x-1)$ and $\hat{x} = 1.39364$. The following diagram shows the graph of f along with the secant.

Returning to the general procedure, the points (x, y) on the secant must satisfy

$$\frac{f(x_2) - f(x_1)}{x_2 - x_1} = \frac{y - f(x_1)}{x - x_1}.$$

Setting $y = 0$ and solving for x, we get

$$x = x_1 - f(x_1)\left[\frac{x_2 - x_1}{f(x_2) - f(x_1)}\right]. \tag{1.4.1}$$

Equation (1.4.1) is the operative statement in the sense that any program that implements the secant method must include this statement. The basic

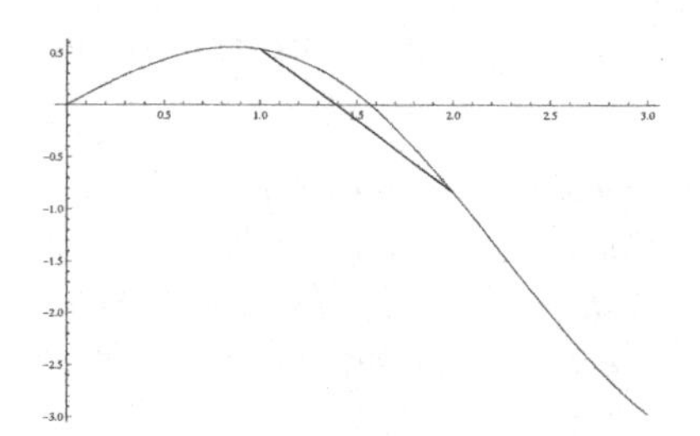

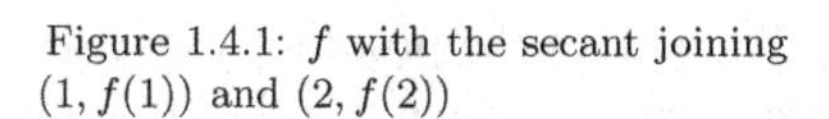
Figure 1.4.1: f with the secant joining $(1, f(1))$ and $(2, f(2))$

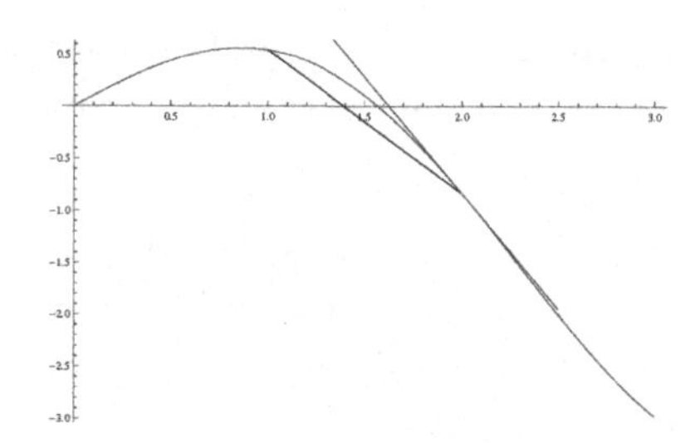

Figure 1.4.2: f concave down near the root

structure of a program for this method is the same as the program for Newton's method. In large measure, we need only to replace (1.3.1) for Newton's method with (1.4.1) for the secant method.

However, there is one wrinkle. In Newton's method we have a single seed which is replaced at each iteration by the current approximate root. For the secant method there are two seeds, x_1 and x_2. When you compute x using (1.4.1) we must replace one of seeds before you compute the next x. In this case, you will need to evaluate f at the three points, x, x_1 and x_2. We know that $f(x_1)$ and $f(x_2)$ have different signs. If $f(x)$ and $f(x_1)$ have the same sign, then replace x_1, if they have different sign (hence, $f(x)$ and $f(x_2)$ have the same sign), then replace x_2.

The secant method is in fact an approximate form of Newton's method. You see this by rewriting (1.4.1) as

$$x = x_1 - \frac{f(x_1)}{\frac{f(x_2)-f(x_1)}{x_2-x_1}}. \tag{1.4.2}$$

If $|x_2 - x_1|$ is small, then the denominator on the right hand side, $(f(x_2) - f(x_1))/(x_2 - x_1)$ is very near to $f'(x_1)$. As the iterative process proceeds, we should expect the successive values for x_1 and x_2 to converge together. In this case the expression (1.4.2) for the approximate root via the secant method will converge to the expression given in (2.1.1) when using Newton's method.

We turn now to error estimation. Suppose f is decreasing and concave down in the interval $[x_1, x_2]$, as is the case for the current example $f(x) = x\cos(x)$. Let $\hat{x}$ denote the approximate root derived from Newton's method and let $\tilde{x}$ denote the approximate root derived from the secant method. It is easy to see that $\hat{x} \leq \tilde{x}$ and that the actual root must lie between.

Figure 2.4.2 illustrates this for the example case. We state this result formally in the following theorem.

Theorem 1.4.1. *Consider a twice differentiable real valued function f defined on an interval $[a, b]$. Suppose that f has a root at $\bar{x}$ in the interval but no relative extrema or inflection points, then the following holds. Each Newton's method estimate $\hat{x}$ and secant method estimate $\tilde{x}$, satisfies $\hat{x} \leq \bar{x} \leq \tilde{x}$ or $\tilde{x} \leq \bar{x} \leq \hat{x}$.*

Proof. There are four cases to consider. We will do the proof for the case when f is decreasing and concave up. If $a < x_1 < \bar{x}$, then $f(x_1) > 0$ and $\hat{x} = x_1 - f(x_1)/f'(x_1) > x_1$. We denote the tangent for f at x_1 by h. Since f is concave up, then h decreases faster than f in the interval $[x_1, \bar{x}]$. Therefore, $\hat{x} \leq \bar{x}$.

On the other side, given seeds $x_1 < x_2$, we know that $f(x_1) > 0$, $f(x_2) < 0$. By definition, the secant estimate $\tilde{x}$ is no larger than x_2. In particular, 1.4.1 may be rewritten as

$$\tilde{x} = x_1 - \frac{f(x_1)}{\frac{f(x_2)-f(x_1)}{x_2-x_1}},$$

by interchanging the role x_1 and x_2. Next, $f(x_2) < 0$, $x_2 - x_1 > 0$ and $f(x_2) - f(x_1) > 0$. Therefore, $\tilde{x} < x_2$.

Let k denote the secant. Since, f is concave up, then for each $x \in [x_1, x_2]$, $f(x) < k(x)$. Therefore, $f(\tilde{x}) < 0$. It follows that $\bar{x} \leq \tilde{x} \leq x_2$. This completes the proof for this case. □

Next, we derive our first example of an error estimating procedure. In particular, $|\hat{x} - \bar{x}| \leq |\tilde{x} - \hat{x}|$. Similarly, $|\tilde{x} - \bar{x}| \leq |\tilde{x} - \hat{x}|$. Hence, $|\tilde{x} - \hat{x}|$ is an upper bound on the absolute error. In other words, we may use $|\tilde{x} - \hat{x}|$ to estimate the error without knowledge of $\bar{x}$.

Theorem 1.4.2. *Suppose that f is twice differentiable function on $[a, b]$ with a root in the interval. Furthermore, suppose that f has no extrema or inflection points in the interval. If $\hat{x}$ is the Newton's method estimate of the root and $\tilde{x}$ is the secant method estimate, then the absolute error for either procedure is bounded by $|\hat{x} - \tilde{x}|$.*

The secant method is also implemented in *Mathematica* via the *FindRoot* command . For the secant method, we need to supply two seeds. For instance, we know that $f(x) = x\cos(x)$ has a root between $x = 1$ and 2. The following *Mathematica* statement will implement the secant method to return an approximate value for the root.

```
x-root = FindRoot[ x Cos(x) == 0, {x, 1, 2} ];
```

Again when using this option of the FindRoot statement, you have access to the iteration count and kick out threshold.

Exercises:

1. Consider $f(x) = (x-1)^2 - 2$ which has a root at $1+\sqrt{2}$. Use the secant method with seeds at 2 and 3 to estimate this root.

2. Execute Newton's method for the function in Exercise 1. Compute the error estimate and verify that actual absolute error is less than the estimated absolute error.

3. Consider the function $f(x) = xe^{-x} - 0.16064$.

a. Use FindRoot to locate a root between $x = 2$ and $x = 3$.

b. Write a program in *Mathematica* implementing the secant method. Use your program to locate a root near $x = 3$. Set the maximal number of iterations to 100 and the kick out threshold to 10^{-5}.

c. How many iterations does your program actually execute before it stops? How does the secant method compare to Newton's method?

d. Use the result of Exercise 1.b from the previous work on Newton's method along with the result of the prior section to get an upper bound on the absolute error.

4. Complete the proof of Theorem 1.4.1

5. For the example in Exercise 1, use *ListPlot* to display 5 Newton's method estimates along with 5 secant method estimates. Use different symbols for the two plots.

6. Prove that if f has a root in $[a, b]$ and no critical or inflection points in the interval, then the sequence of Newton method approximate roots is a convergent sequence.

7. Repeat Exercise 6 for the secant method.

Chapter 2

Linear Systems and Optimization

Introduction

In this chapter, we develop the basic procedures necessary to solve linear systems of equations and locate functional extrema. We introduce both standard techniques and their extensions for large matrices. This includes methods for finding eigenvalues, solving linear systems of equations and optimization problems. We resolve these questions in the context of a course supported by a 4GL programming language. We have all solved max/min problems and linear systems of equations with pencil and paper. For the most part we do this for carefully selected textbook exercises. You may have found these problems to be tedious. Now, we find that *Mathematica* makes these problems effortless and routine. Even in cases with thousands of unknowns or even hundreds of thousands of unknowns.

We begin with the *Mathematica* implementation of Gauss-Jordan elimination. Next, we consider techniques for solving large linear systems. In this context we develop Jacobi and Gauss-Seidel methods as special cases of the Richardson method. We follow this with the power method for finding eigenvalues and eigenvectors. At the same time, we introduce Krylov subspaces including a brief discussion of CG or conjugate gradient method.

Our point of view is to provide an introduction to some of the ideas that are necessary to deal with large data sets. An important aspect of this development is the operator norm introduced in the second section. This result is arguably one of the two most important of this book. We encounter the other in the next chapter.

During this development, we have occasion to apply the Gelfand theorem on the spectral radius of a matrix. We will reference this result without proof as the usual argument requires the Jordan canonical form of a matrix.

We end the chapter considering max/min problems in several variables. Along the way we show that finding the roots of a function of several variables may be recast as a max/min problem. In addition, these procedures can be used to solve linear systems with singular coefficient matrix. Both questions relate to the root finding techniques introduced in the prior chapter.

2.1 Linear Systems of Equations

Here we develop procedures for solving a linear system of equations with non-singular coefficient matrix. The case for singular coefficient matrices will be presented later in the chapter.

First we need to consider the program code necessary to do matrix arithmetic. The following statements define and display a four dimensional column vector, v.

```
v = {1,2,3,4} ;
Print[MatrixForm[v]];
```

The following two statements define 4 by 4 matrices A and B. The following statement multiplies the two matrices, then prints the product, C, formatted as a matrix. The final step multiplies A times v to get the 4-tuple w. Next, we multiply w by a scalar 5. Notice that we use a period to multiply matrices and matrices times vectors whereas we use the * to multiply scalars and vectors. Even though the entries of these matrices and vectors are integers we use at least one decimal point to ensure all processing will be decimal processing.

```
A = {{1.,2,3,4},{2,3,4,5},{5,4,3,2},{4,3,2,1}};
B = {{0.,1,0,0},{1,0,0,0},{0,0,0,1},{0,0,1,0}};
C = A.B;
v = {1.,2,3,4};
Print[MatrixForm[C]];
w = A.v;
w = 5*w;
```

Some other useful linear algebra functions are $Transpose[A]$, $Transpose[v]$, $Inverse[A]$, $IdentityMatrix[n]$, $Norm[v]$ and $Length[v]$. The first one produces the transpose of the matrix A. The next one changes the column vector v into a row vector. The next two produce the inverse of

A, and the $n \times n$ identity matrix. $Norm[v]$ returns the length of the vector v while *Length*$[v]$ returns 4 if v is a four-tuple. Finally, $A[[i]][[j]] = A[[i,j]]$ returns the ij^{th} entry of A and $v[[i]]$ is the ith entry of the vector v.

Consider a linear system $Ax = b$, where A is a non-singular $n \times n$ matrix and x, $b \in \mathbb{R}^n$. The condition that A is non-singular assures us that the system has a unique solution given by $x = A^{-1}b$. In addition, we know that A is non-singular if and only if A is row equivalent to the $n \times n$ identity, I_n.

The standard process of solving a linear system of equations is called *Gauss-Jordan elimination*. This method is implemented in *Mathematica*. Suppose you have the linear system

$$\begin{pmatrix} 1 & 2 & 3 \\ 4 & 5 & 6 \\ 7 & 8 & 9 \end{pmatrix} \begin{pmatrix} x_1 \\ x_2 \\ x_3 \end{pmatrix} = \begin{pmatrix} 1 \\ 2 \\ 3 \end{pmatrix}$$

In order to solve this system in *Mathematica*, you define the coefficient matrix and constant vector via

```
coefMat = {{1,2,3},{4,5,6},{7,8,9}};
conVec = {1,2,3};
```

And then solve the system via the statement.

```
solVec = LinearSolve[coefMat, conVec];
```

Along with Gauss-Jordan elimination (*LinearSolve*), *Mathematica* will also return the *LUDecomposition*. In order to understand what *Mathematica* does when solving a linear system, we need to look at LUDecomposition. At this stage, you need to recall a number of things about *row equivalence* and *elementary row operations* as they apply to linear systems of equations.

(1) Matrices A and B are *row equivalent* provided there exist *elementary matrices* $E_1, ..., E_m$ with $B = (\prod_{i=1}^{m} E_i)A$.
(2) Multiplication on the left by an elementary matrix implements the corresponding *elementary row operation.*
(3) There are three elementary row operations.
(4) The *type-1 elementary row operation* exchanges two rows of a matrix. It is denoted by $E_{(i,j)}$, indicating that rows i and j are exchanged.
(5) The *type-2 elementary row operation* multiplies a row by a nonzero scalar. It is denoted by $E_{\alpha(i)}$, indicating that row i is multiplied by α.

(6) The *type-3 elementary row operation* adds a scalar times one row to another. It is denoted by $E_{\alpha(i)+j}$, indicating that α times row i is added to row j. In addition, for each type, the inverse is of the same type.

(7) All elementary matrices are non-singular. Their inverses are given by $E_{(i,j)}^{-1} = E_{(i,j)}$. $E_{\alpha(i)}^{-1} = E_{\alpha^{-1}(i)}$ and $E_{\alpha(i)+j}^{-1} = E_{-\alpha(i)+j}$. Notice that the inverse of an elementary matrix is the elementary matrix that reverses the operation.

(8) If $Ax = b$ and $Bx = c$ are linear systems and D is a product of elementary matrices then $B = DA$, $c = Db$ implies that the two systems have the same solution. These systems are called *equivalent systems.*

Gauss-Jordan elimination hinges on the fact that if A were upper triangular, then the system, $Ax = b$ could be easily solved. Indeed, the row-echelon form of A is upper triangular. The U in LU-Decomposition refers to this upper triangular matrix. The problem of solving the linear system reduces to finding an upper triangular matrix U, row equivalent to A. In this case, there is a vector c with $Ax = b$ and $Ux = c$ equivalent.

Suppose $A = [\alpha_{i,j}]$ and that $\alpha_{1,1} \neq 0$. Then it is a simple matter to clear out the entries of the first column of A below the $1,1$ entry. For this purpose we only need elementary operations of type-3. The corresponding elementary matrices, $E_{\alpha(1)+j}$ are all lower triangular since each $j > 1$. Since the product of lower triangular matrices is lower triangular, then we have a lower triangular matrix L_1 with $L_1 A = A_1 = [\hat{\alpha}_{i,j}]$ where $\hat{\alpha}_{i,1} = 0$ for all entries of column 1 below the $1,1$ entry. Furthermore, L_1 is nonsingular.

Now, we move to the second column. If $\hat{\alpha}_{(2,2)} \neq 0$, then we can clear out below the $2,2$ position using only type-3 elementary operations. As before, there is a lower triangular matrix L_2 with $L_2 L_1 A = L_2 A_1 = A_2$. As before, $L_2 L_1$ is lower triangular and nonsingular. If the $3,3$ entry of A_3 is non-zero, then the process continues. Indeed, if at each step the next diagonal entry is not zero, then we clear out below the diagonal element. At the end, we have

$$(L_n ... L_2 L_1)A = U \tag{2.1.1}$$

where each L_i is lower triangular and nonsingular and U is upper triangular. Multiplying (2.1.1) by $(L_n ... L_1)^{-1} = L_1^{-1} ... L_n^{-1} = L$ yields the relation $A = LU$, where L is lower triangular (the inverse of a lower triangular matrix is lower triangular) and U is upper triangular. This is the LU-decomposition for A. As long as we do not encounter a zero on the main diagonal, the process is as described.

Now, we only need consider what to do if there is a diagonal entry that is zero. In this case, if the i, i entry is zero, there must be a non-zero entry below the diagonal. If this were not the case then A would be singular. See Exercise 2. If we begin with

$$\begin{pmatrix} 1 & 2 & 3 \\ 4 & 8 & 6 \\ 7 & 5 & 9 \end{pmatrix},$$

then upon clearing the first column, we have

$$\begin{pmatrix} 1 & 2 & 3 \\ 0 & 0 & -6 \\ 0 & -9 & -12 \end{pmatrix}.$$

After applying a type-1 elementary operation, $E_{(2,3)}$, we have the upper triangular matrix

$$\begin{pmatrix} 1 & 2 & 3 \\ 0 & -9 & -12 \\ 0 & 0 & -6 \end{pmatrix} = E_{(2,3)}(E_{-7(1)+3}E_{-4(1)+2})A.$$

In particular, whenever we encounter a zero on the diagonal, then we must introduce a type-1 elementary matrix which interchanges rows to bring a nonzero entry to the diagonal.

Looking more closely at the product of a type-1 and a type-3,

$$E_{(i,j)}E_{\alpha(s)+t} = E_{\alpha(s)+t}E_{(i,j)} \tag{2.1.2}$$

provided s, t are not elements of the set $\{i, j\}$. Otherwise

$$E_{(i,j)}E_{\alpha(i)+t} = E_{\alpha(j)+t}E_{(i,j)} \tag{2.1.3}$$

and

$$E_{(i,j)}E_{\alpha(s)+i} = E_{\alpha(s)+j}E_{(i,j)}. \tag{2.1.4}$$

We know from the context that $s < i < j$. Indeed, i is the current column whereas s references the source row at a prior step. Since we are interchanging a lower row for an upper, then $i < j$. The first thing to conclude is that the case described in (2.1.3) cannot occur in our setting. For the remaining two cases, the type-3 matrices are all lower triangular.

For Gauss-Jordan elimination, suppose we have arrived at a state,

$$(L_m...L_1)A = A_m$$

and suppose that the i, i entry is zero. If the (i, j) entry, $i > j$, is nonzero, then we apply $E_{(i,j)}$ to both sides and continue to clear down. This will yield

$$L_k L_{k-1} ... L_{m+1} E_{(i,j)} (L_m ... L_1) A = A_{m+k}.$$

Using (2.1.2) and (2.1.4.) there are lower triangular matrices K_i with

$$K_k K_{k-1} ... K_1 E_{(i,j)} A = A_{m+k}.$$

Continuing in this manner we come the final state or revision of (2.1.1),

$$L_n ... L_2 L_1 P_m ... P_2 P_1 A = U. \tag{2.1.5}$$

Hence, we may restate (2.1.1) as

Theorem 2.1.1. *If A is a nonsingular $n \times n$ matrix, then there exists a lower triangular matrix L, an upper triangular matrix U and a nonsingular matrix P with $PA = LU$. The upper triangular matrix U is the row echelon form of A, L is a product of type-3 elementary matrices and P is a product of type-1 elementary matrices.*

The matrix, P, which is the product of type-1 matrices, is called a *permutation* matrix. The idea here is simple. Each type-1 elementary operation transposes two rows of the matrix A. In turn, the product iterates transpositions into a permutation of the rows. Hence, the term, permutation matrix.

For any nonsingular matrix A, this information is available via the *Mathematica* function *LUDecomposition*. For instance, for the matrix,

$$A = \begin{pmatrix} 1 & 2 & 3 \\ 4 & 8 & 6 \\ 7 & 5 & 9 \end{pmatrix}$$

```
LUDecomposition[A]
```

will return the following output.

```
{{{1,2,3},{7,-9,-12},{4,0,-6}},{1,3,2},1}.
```

The first item in the list is a 3×3 matrix. The upper triangular part of this matrix is U. The lower triangular part (after placing 1's along the diagonal) is L. Now we know that

$$U = \begin{pmatrix} 1 & 2 & 3 \\ 0 & -9 & -12 \\ 0 & 0 & -6 \end{pmatrix}, L = \begin{pmatrix} 1 & 0 & 0 \\ 7 & 1 & 0 \\ 4 & 0 & 1 \end{pmatrix}.$$

Next $\{1,3,2\}$ is interpreted as the permutation that sends $1 \to 1$, $2 \to 3$, $3 \to 2$. Hence,

$$P = \begin{pmatrix} 1 & 0 & 0 \\ 0 & 0 & 1 \\ 0 & 1 & 0 \end{pmatrix}.$$

The final entry in the output is called the *condition number.* This number is useful when estimating the error. We develop the condition number in the following section.

Exercises:

1. Apply LUDecompostion to the following matrices. Write out L, U and P.

a. $\begin{pmatrix} 1. & 1. & 0 \\ 1. & 1. & 3. \\ 0. & 1. & -1. \end{pmatrix}$

b. $\begin{pmatrix} 1. & 2. & 1. & 7. \\ 2. & 0. & 1. & 4. \\ 1. & 0. & 2. & 5. \\ 1. & 2. & 3. & 11. \end{pmatrix}$

c. $\begin{pmatrix} 1. & 2. & 3. \\ 1. & 1. & 1. \\ 5. & 7. & 9. \end{pmatrix}$

2. Let $A = [\alpha_{i,j}]$ be an $n \times n$ matrix with real entries. Suppose that there is an m with $\alpha_{i,j} = 0$ for $i \geq m$, $j \leq m$ and $\alpha_{i,i} \neq 0$ for $1 \leq i < m$. Prove that A is singular.

3. The *Mathematica* statement *Eigensystem[A]* returns n+1 vectors for an $n \times n$ matrix A. The entries of the first vector are the eigenvalues of A. The remaining vectors are the corresponding eigenvectors. Apply Eigensystem to the matrices listed in (1). Recall that a matrix is singular if it has zero as an eigenvalue. Also, when looking at computer output a number close to zero should be considered zero. Which of the matrices in (1) are singular?

4. Prove the following for an $n \times n$ triangular matrix,

a. The product of upper (lower) triangular matrices is upper (lower) triangular,

b. The inverse of a triangular matrix is a triangular matrix of the same type.

5. Build a 500×500 matrix $A = [\alpha_{i,j}]$ with diagonal entries $\alpha_{i,i} = 1$, super diagonal entries $\alpha_{i,i-1} = -0.5$ and subdiagonal entries $\alpha_{i+1,i} = -0.5$. (Hint: $A = IdentityMatrix[500] - IdentityMatrix[500]$ will return a zero matrix of the desired size. From this point you only need to load $A_{i,j} = 1$ for each i and so forth.)

a. Execute *LinearSolve* for the linear system $Ax = b$ where $b_i = 1$ for every i. Use the *TimeUsed* function to determine the required CPU time.

b. Repeat Exercise *a* for a 100×100, ..., 400×400. Plot CPU time against the matrix size.

6. In a forest two species of tree dominate, T_A and T_B. Tree T_A is endangered and rumored to be on the edge of extinction. The forest manager is charged with determining the distribution of the two trees in the forest over the coming years. From a recent inventory carried out for over an extended period, the following data is known.

a. In a given year the manager can expect 1 percent of T_A to die and 5 pecent of T_B.

b. There is a 25 percent chance that a dead tree is replaced by the endangered T_A and a 75 percent chance that it will be replaced by T_B.

c. All dead trees are replaced by one or the other type of tree. New trees enter the forest only by replacement.

d. Currently, there are 10 T_A and 990 T_B.

e. Since all collected data is annual, the manager sets the unit of time τ to be 1 year. Note that $t_A(\tau) = total(T_A)$ at a given time, and $t_B(\tau)$ is defomed similarly.

To solve the problem the manager has done the following

For a given year, the number of dead trees is $(0.01)t_A(\tau) + (0.05)t_B(\tau)$. The number of replacements from the type T_A is

$$(0.25)(0.01)t_A(\tau) + (0.25)(0.05)t_B(\tau).$$

The type T_B replacements is

$$(0.75)(0.01)t_A(\tau) + (0.75)(0.05)t_B(\tau).$$

Unfortunately the manager has resigned leaving you to finish the study. He has left no contact information.

a. Write $t_A(\tau + 1)$ in terms of $t_A(\tau)$ and $t_B(\tau)$. (Hint: What factors determine $t_A(\tau + 1)$. These are: no T_A dies or there are deaths of either type that are replaced by T_A or there are deaths of T_A that are replaced by T_B. Which factors are positive, which are negative?)

b. Write $t_B(\tau+1)$ in terms of $t_A(\tau)$ and $t_B(\tau)$.

c. Recast the results of a and b in matrix form; that is find A such that

$$\begin{pmatrix} t_A(\tau+1) \\ t_B(\tau+1) \end{pmatrix} = A \begin{pmatrix} t_A(\tau) \\ t_B(\tau) \end{pmatrix}.$$

d. Use *Eigensystem*[A] to find the eigenvectors and eigenvalues of A. Confirm that A has a diagonal representation. We next recast the problem for a diagonal of A.

e. Write

$$\begin{pmatrix} t_A(0) \\ t_B(0) \end{pmatrix}$$

as a linear combination of the eigenvectors of A.

f. Write

$$\begin{pmatrix} t_A(1) \\ t_B(1) \end{pmatrix}$$

using the result of e and the eigenvalues of A.

g. Use the result of f to write a formula for

$$\begin{pmatrix} t_A(\tau+1) \\ t_B(\tau+1) \end{pmatrix}$$

in terms of $t_A(\tau)$ and $t_B(\tau)$, the eigenvectors and the eigenvalues.

h. Will T_A disappear from the forest? If so, when? Plot t_A over the next 25 years.

i. Is it reasonable to expect that the matrix A will not change in the next 25 years?

2.2 The Norm and Spectral Radius of a Linear Transformation

We begin with the operator norm of the linear transformation. This concept is key to estimating the error in Gauss-Jordan elimination. But that is only the beginning. The operator norm is essential to understanding iterative processes with linear transformations. Hence, it plays a critical role in large matrix processes. Recast into infinite dimensional vector spaces, the linear operators with bounded norm are central. For instance they play a key role in quantum mechanics and in linear partial differential equations.

We will develop the operator norm for real vector spaces. The same arguments work for the complex case.

Recall the norm of a vector in $\mathbb{R}^n$, $\|v\| = (\sum_i v_i^2)^{1/2}$. We extend this idea to a linear operator in the following definition.

Definition 2.2.1. The *operator norm* of a linear transformation A of a finite dimensional vector V is the least constant M which satisfies $\|Av\| \leq M\|v\|$ for any v in V. It is usual to write $\|A\|$ for the operator norm.

Notice that the norm of A as defined here is independent of the matrix representation.

Our initial task is to prove that $\|A\|$ exists. Toward this end, we develop notation. First let $e_1, ..., e_n$ denote the standard basis vectors. Here, e_i is the $n - tuple$ with 1 in position i and zeros elsewhere. If v is a unit vector, then $v = \sum_{i=1}^n \xi_i e_i$ and $1 = \|v\|^2 = \sum_{i=1}^n \xi^2$, since the e_i form an orthonormal frame. Hence, for each i, $|\xi_i| \leq 1$. We calculate

$$\|Av\| = \|A(\sum_{i=1}^n \xi_i e_i)\| = \|\sum_{i=1}^n \xi_i A e_i\| \leq \sum_{i=1}^n |\xi_i| \|Ae_i\|.$$

Taking $m = \max_i \|Ae_i\|$, we have proved that

$$\|Av\| \leq nm, \tag{2.2.1}$$

for the case of a unit vector v. We now prove the existence theorem for the operator norm.

Theorem 2.2.1. *Any linear transformation of a finite dimensional real or complex vector space has finite operator norm.*

Proof. We continue the notation. Given $u \neq 0$ in V, $v = u/\|u\|$ is a unit vector. By (2.2.1), $\|Av\| \leq K$ where $K = mn$. Hence, for nonzero u,

$$\|Au\| = \|u\|\|A(\frac{1}{\|u\|}u)\| = \|u\|\|Av\| \leq \|u\|K. \tag{2.2.2}$$

The norm is the least constant K that satisfies (2.2.2). □

Note that the theorem proves the existence of the norm without providing a means to calculate it. Rather, we are given an upper bound.

In this section, we have already seen two functions that we call a norm, the Euclidean norm of a vector in $\mathbb{R}^n$ and now the norm of A. These are both examples of a more general concept.

Definition 2.2.2. Let V be a real or complex vector space. A *norm* on V is a function $\|.\|$ taking values in $\mathbb{R}$ such that

(1) for any nonzero vector, $\|v\| > 0$,

(2) for any scalar α, $\|\alpha v\| = |\alpha|\|v\|$,
(3) for u and v in V, $\|u + v\| \leq \|u\| + \|v\|$.

In this case V is called a *normed linear space.*

Property (3) is called the *triangle inequality.*

The following theorem states that the operator norm is a norm.

Theorem 2.2.2. *The vector space of linear transformations of the space V with the operator norm is a normed linear space.*

Proof. See Exercise 3. □

A norm gives rise to a metric and with a metric we can talk about convergence, open sets and continuous functions.

Definition 2.2.3. Given a set X, a *metric* is a real valued function $d : X \times X \to \mathbb{R}$ satisfying the following.

(1) For distinct x and y in X, $d(x, y) > 0$ and $d(x, x) = 0$.
(2) For x and y in X, $d(x, y) = d(y, x)$.
(3) For x, y and z, $d(x, y) + d(y, z) \geq d(x, z)$.

The last property is also called the *triangle inequality*. It is not difficult to prove (see Exercise 9) that $\|x - y\| = d(x, y)$ defines a metric on a normed linear space. Before moving on, note that as a consequence of Theorem 2.2.1, linear transformations are continuous. We state this formally in the following theorem.

Theorem 2.2.3. *A linear transformation A defined on a real vector space $\mathbb{R}_n$ is uniformly continuous.*

Proof. Take $\epsilon > 0$ and select $\delta = \epsilon/\|A\|$. By Theorem 2.2.1,

$$\|Au - Av\| = \|A(u - v)\| \leq \|A\|\|u - v\| < \epsilon,$$

provided $\|u - v\| < \delta$. □

We now turn to the error analysis for the solution to a linear system. Consider the linear system $Av = b$. We denote the solution by v and the computed solution $\hat{v}$. In turn, we set $\hat{b} = A\hat{v}$. Now, we compute the relative normed error,

$$\frac{\|v - \hat{v}\|}{\|v\|} = \frac{\|A^{-1}(b - \hat{b})\|}{\|v\|} \leq \|A^{-1}\|\frac{\|b - \hat{b}\|}{\|v\|}.$$

Next, we multiply top and bottom by $||b||$ and use $||b|| \leq ||A|| ||v||$,

$$\frac{||v - \hat{v}||}{||v||} \leq ||A^{-1}|| \frac{||b - \hat{b}|| ||b||}{||v|| ||b||}$$

$$\leq ||A|| ||A^{-1}|| \frac{||b - \hat{b}|| ||v||}{||v|| ||b||} = ||A|| ||A^{-1}|| \frac{||b - \hat{b}||}{||b||}.$$

The right hand expression is called the *relative normed residual* and the coefficient $C = ||A|| ||A^{-1}||$ is the *condition number*.

We state this result formally as a theorem.

Theorem 2.2.4. *For the linear system $Av = b$, the relative normed error is bounded by the condition number times the relative normed residual.*

There is no problem to compute the relative normed residual. The condition number (equivalently the operator norm) is another story. It is related to the eigenvalues of A. We investigate this relationship in the exercises.

If the condition number is very large, then the coefficient matrix for the system is singular or nearly singular and the results returned by LinearSolve are not considered reliable. When this is the case, *Mathematica* will return a warning that the coefficient matrix is *ill conditioned.* But all of this presupposes that the matrices are not too large. For large matrices we are on our own. Indeed, large nonsingular matrices may have very large condition numbers. See Exercise 7.

Because of Theorem 2.2.3, we need to develop the operator norm in order to estimate the condition number. To begin, note that $||A|| ||B|| \geq ||AB||$. And there are cases where equality fails. (See Exercises 2, 4 and 5.) Therefore, in order to know the condition number, you must know the operator norm. Before proceeding, we need some terminology.

Definition 2.2.4. For a real or complex $n \times n$ matrix A, the *spectral radius* is the $\max_i |\lambda_i|$ where λ_i, $i = 1, 2, ..., n$ are the eigenvalues of A. The spectral radius of A is denoted $\sigma(A)$.

Alternatively, we write $\mu(A)$ for the absolute value of the smallest eigenvalue. The proof of the following theorem is included in the exercises.

Theorem 2.2.5. *Let A be an $n \times n$ real nonsingular matrix. Then the following hold.*

i. If λ is an eigenvalue for A, then $|\lambda| \leq ||A||$.
ii. In particular, $\sigma(A) \leq ||A||$ and $1/\mu(A) \leq ||A^{-1}||$.
iii. $\sigma(A)/\mu(A) \leq C$, the condition number for A.

Proof. See the Exercise 4. □

For the case of a real symmetric matrix the situation is much more tractable. Indeed, it is a happy circumstance that linear systems with symmetric coefficient matrices arise naturally. In particular, this is the case for processes such as diffusion. The key here is that a real symmetric matrix has n orthonormal eigenvectors.

Theorem 2.2.6. *Let A be an $n \times n$ real symmetric matrix, then the following hold.*

i. The norm of A is equal to the spectral radius.

ii. $||A^k|| = ||A||^k$ for any positive integer k.

Proof. See Exercise 8. □

There is one more result that will be useful as we go forward. The proof uses the Jordan canonical form. We state the theorem here without proof. It is attributed to Gelfand. [Loustau (2016)]

Theorem 2.2.7. *For any $n \times n$ complex matrix,*

$$\lim_k ||A^k||^{1/k} = \sigma(A). \tag{2.2.3}$$

Exercises:

1. Determine which of the matrices in Exercise 1 of Section 1 are ill conditioned.

2. Prove that for real or complex matrices, $||A|| ||B|| \geq ||AB||$.

3. Prove Theorem 2.2.2. (Hint: For part three consider $||(A+B)v|| = ||Av + Bv|| \leq ||Av|| + ||Bv||$).

4. The following exercises lead to an estimator for the condition number of a nonsingular matrix A.

a. Let v be an eigenvector for A with eigenvalue λ. Prove that $||Av|| = |\lambda| ||v||$. Prove that $|\lambda| \leq ||A||$.

b. Prove that λ is an eigenvalue for A if and only if $1/\lambda$ is an eigenvalue for A^{-1}. (Hint: Consider $v = \lambda A v$ and multiply by A^{-1}.)

c. Let $\mu(A)$ be the absolute value of the smallest eigenvalue of A. Prove that $\sigma(A) \leq ||A||$ and $1/\mu(A) \leq ||A^{-1}||$.

d. Prove that $\sigma(A)/\mu(A) \leq C$, the condition number of A.

5. Prove that $\|A\| > 1$ while $\sigma(A) = 1$, where

$$A = \begin{pmatrix} 1 & 1 \\ 0 & 1 \end{pmatrix}.$$

6. Repeat Exercise 5 of Section 2.1. But now execute *LU decomposition.* Look at the condition number as a function of the matrix size. What is happening? Are these matrices nonsingular? Do they have a zero eigenvalue? The *Eigenvalues* function in *Mathematica* will help here.

7. For the matrices in Exercise 6, set the first row to $(1, 0, 0, ..., 0)$ and the last row to $(0, 0, ..., 1)$. Check the condition number. Also use *Eigenvalues* to verify that the matrices are nonsingular.

8. Prove Theorem 2.2.5 as follows. Suppose that A is $n \times n$ real symmetric with orthonormal eigenvectors $v_1, ..., v_n$.

a. Let v be a unit vector with $v = \sum_i \xi_i v_i$ where the v_i are orthonormal eigenvectors. Prove that $\sum_i \xi_i^2 = 1$.

b. Prove that $\sigma(A) \geq \|A\|$, conclude that $\sigma(A) = \|A\|$.

c. Prove that λ is an eigenvalue for A if and only if λ^k is an eigenvalue for A^k.

d. Prove that $\|A^k\| = \|A\|^k$.

9. Prove that given a normed linear space, $d(u, v) = \|u - v\|$ defines a metric.

2.3 Large Matrix Techniques

In this section we consider techniques for solving the linear system $Ax = b$ when A is simply too large to complete a Gauss-Jordan elimination. Stated otherwise, we expect that A is nonsingular but it is too large to compute A^{-1}. This does not necessarily mean that the matrix is too large to process. Recall that Gauss-Jordan elimination accumulates round off error at the lower right. Hence, the larger the matrix, the less reliable the process. In turn, the unreliability is reflected in the condition number. See Exercise 5 and 6 of Section 2.2.

We begin with a paradigm due to Richardson referred to as the *residual correction* method. It best summarizes large matrix processing.

We begin with a guess for the inverse of A. If B is our estimated inverse for A, then set $x_1 = Bb$, $e_1 = x_1 - x$ and $r_1 = b - Ax_1$, the first approximate solution, first error and first residual. Continuing, we have the $(k+1)^{st}$ solution, $x_{k+1} = x_k + Br_k$, error, $e_{k+1} = x - x_{k+1}$ and residual, $r_{k+1} = b - Ax_{k+1}$. If we expand the $(k+1)^{st}$ error we have

$$\begin{aligned} e_{k+1} &= x - x_{k+1} = x - x_k - Br_k = x - x_k - B(b - Ax_k) \\ &= e_k - Bb + BAx_k = (I - BA)e_k = (I - BA)^k e_1. \end{aligned}$$

The idea behind the Richardson procedure is now apparent. If B is a good estimate for A^{-1} then we expect $I - BA$ to be small, in the sense that its norm is less than one. In this case, we can prove that $x_k \to x$. Indeed, it suffices to prove that $e_k \to 0$.

$$\|e_k\| = \|(I - BA)^{k-1}e_1\| \leq \|I - BA\|^{k-1}\|e_1\| \to 0,$$

using Exercise 1 of the prior section. However, we can do better.

Theorem 2.3.1. *Consider the linear system $Ax = b$ and suppose that B is an approximate inverse for A such that $\sigma(I - BA) \leq 1$, then there is an integer j such that the subsequence $x_{nk+1} \to x$ for any $k > j$. Furthermore, any two convergent subsequences have the same limit.*

Proof. By Theorem 2.2.5, $\lim_k \|(I - BA)^k\|^{1/k} < 1$. Therefore, there is an integer j such that $\|(I - BA)^k]\| < 1$ for every $k \geq j$. Indeed, the kth root of a positive number is less than 1 only if the number is less than 1. Now, if $k > j$, then $\|e_{nk+1}\| \leq \|(I - BA)^{nk}\|\|e_1\| \leq \|(I - BA)^k\|^n\|e_1\| \to 0$ as $n \to \infty$. The final assertion holds as any two convergent subsequences have a common convergent subsequence. □

As consequence of Theorem 2.3.1, the spectral radius of $I - BA$ is critical. At the end of this section we introduce a technique to estimate the spectral radius of a matrix.

We look at two specific techniques. Both are based on Richardson. The first is credited to Jacobi. Continuing with the notation of the theorem, suppose that the diagonal entries $\alpha_{i,i}$ of A are not zero. Then set B so that $B_{i,j} = 0$, $i \neq j$ and $B_{i,j} = 1/\alpha_{i,i}$. Most often the *Jacobi method* is used when A is diagonally dominant, but there is no necessity for that.

As an example, consider the following. Suppose A is a 50×50 matrix. This is so small that there is no difficulty computing the actual solution via Gauss-Jordan elimination, and then the actual error for the Jacobi process.

For instance, set A to be tridiagonal matrix with entries $\alpha_{i,i} = 0.5$ on the diagonal, $\alpha_{i,i-1} = \alpha_{i,i+1} = 0.25$ on the super diagonal and sub diagonal and zeros elsewhere. Next, take b to have i^{th} entry equal to $1/i$. The spectral radius if $I - BA$ is approximately $0.998 < 1$. We expect the Jacobi method to converge slowly. Indeed, this is the case. Starting with $x_0 = (1, 1, ..., 1)$ we have relative normed error of 1.101438 after 50 iterations, 0.41144 after 500 iterations and 8×10^{-5} at 5,000 iterations.

Another form of the Richardson paradigm is the *Gauss-Seidel method.* In general, it is much faster than the Jacobi. In this case we consider L to be the lower triangular part of A with the diagonal included. It is easy to invert L. Indeed, if the diagonal entries of A are nonzero, then a sequence of type 3 elementary operations will yield a row equivalent diagonal matrix. We follow these with type 2 operation to get the identity matrix. Now executing the same operations on the identity matrix produces the inverse of $L^{-1} = B$.

Repeating the prior example for this case, the spectral radius of $I - BA$ is 0.996. Nearly the same as for the Jacobi method. However, this time the relative normed error is 0.6438 for 50 iterations, 0.1102 for 500 and 4×10^{-9} for 5,000. These results are orders of magnitude better. It is not surprising that the Gauss-Seidel method is preferred.

We need to be careful when implementing these techniques. If we could execute ordinary matrix products, then we would be able to execute Gauss-Jordan elimination and arrive at the solution of the linear system in the usual manner. Given we cannot compute BA using ordinary procedures we must take advantage of the probable sparse nature of the matrices. For instance, B is a diagonal matrix in one case and lower triangular in the other. In the first case $BA = C$ where each $C_{i,j} = B_{i,i}A_{i,j}$ and in the Gauss-Seidel case, $C_{i,j} = \sum_{k=1}^{i} B_{i,k}A_{k,j}$. In each case it is better to write the code directly, rather than depend on a general purpose matrix multiplication. Continuing, we take advantage of the zero entries of $(I - BA)$ by computing $(I - BA)^k$ as $(I - BA)(I - BA)^{k-1}$.

Of course these iterative processes require a stop criterion. There are two standard procedures. The first, stops the iteration when the relative normed residual, $\|b - Ax_n\|/\|b\|$, is smaller than some predetermined threshold. The second, applies the threshold to the relative normed difference between two successive solution estimates, $\|x_n - x_{n+1}\|/\|x_n\|$.

In another direction, we want to estimate $\sigma(A)$ for a large matrix A. There is a procedure to estimate the spectral radius of a real matrix. It is called the *power method.*. The method is remarkably simple. We begin with an initial estimate for an eigenvector, x_0. We iteratively calculate $x_{n+1} =$

$Ax_n/\|Ax_n\|$. As with most anything else the method is not universally applicable. With the following theorem we state applicability criteria for the power iteration method. We omit the proof as the theorem requires the Jordan canonical form. We note that the proof is similar to the one for the Gelfand result (Theorem 2.2.7).

Theorem 2.3.2. *Let A be an $n \times n$ matrix with spectral radius $\sigma(A)$ associated to eigenvalue λ and eigenvector x_λ. Suppose that λ is not the eigenvalue for two independent eigenvectors. If x_0 is a vector with nonzero x_λ component, then the sequence $\{x_n : x_n = Ax_{n-1}/\|Ax_{n-1}\|\}$ converges to x_λ and $\{Ax_n\}$ converges to λx_λ.*

In the exercises, we develop a proof for the case of a symmetric matrix A.

We end this section with a brief discussion of *Krylov* subspaces. The idea behind Krylov subspaces is to generalize the idea of an iterative matrix multiplication process. This turns out to be useful as there are very efficient linear solve procedures that are best developed with Krylov subspaces. Conjugate gradient is one example. The basic idea is that the sequence of approximate solutions $\{x_j\}$ is a subset of a finite dimensional vector space.

First suppose that $\|I - BA\| < 1$, in other words, suppose that the sequence of approximate solutions is convergent. Next, we take k so that x_k is a linear combination of x_0, ..., x_{k-1}. Since the underlying vector spaces are all finite dimensional, then we are assured that k exists. We set W equal to the linear span of x_0, ..., x_{k-1} and claim that W is BA invariant. To verify the claim, notice that $BAx_{j-1} = x_j - x_{j-1}$. If $j < k-1$, then $x_j - x_{j-1} \in W$ and it follows that $BAx_{j-1} \in W$. For the case $j = k-1$, x_k is in W. Hence, $BAx_{k-1} = x_k - x_{k-1}$ lies in W.

Next, we look at $x_{k+1} = x_k + BAx_k$. This is also an element of W. In fact, we can now conclude that $x_j \in W$ for each j. In other words, the entire iterative process is played out inside of W. In turn, it follows that x lies in W. This is not difficult to verify since W is finite dimensional and hence closed. Thus, x, the limit of the iteration, is also an element of W.

Suppose $k = dimW$, then the solution to the linear system is a combination of k known vectors. Hence, what was an $n \times n$ problem is now a $k \times k$ problem. Depending on the setting, this may be significant. In the literature, W is called the kth Krylov subspace. We look at a particular case in Exercise 8.

However, this is not the end of the story. Since we are computing elements of a sequence that converges to the solution, then $\|x_n - x_{n+1}\| \to 0$.

Hence, it may well be that x_n and x_{n+1} are independent but due to round off error, they appear dependent. We get around the problem by replacing each x_n with a vector orthogonal to all x_m, $m < n$. In the literature, this is done with a clever modification of the Gram-Schmidt process. Given the orthonormal frames for W, we have the basis needed to solve the new linear system.

Exercises:

1. Consider the example with the 50×50 tridiagonal coefficient matrix from the text. Execute the Jacobi method using the stop threshold 10^{-5}.

2. Apply the Gauss-Seidel method to the 50×50 tridiagonal matrix from the text. Build a corresponding 100×100 tridiagonal matrix and apply Gauss-Seidel. How does the size change affect the relative normed error at 50, 500 and 5000 iterations.

3. Repeat Exercise 2 using an exit threshold value of 10^{-5}. Compare relative normed residual exit test to the relative step difference exit test. Which one exits sooner?

4. Execute Gauss-Seidel for several symmetric matrices of various sizes. Use *Random* in *Mathematica* to generate the matrices. Set the exit threshold to 10^{-5}. Determine the approximate time and iteration counts for each example.

5. Let L be a lower triangular matrix with inverse B. Write an expression for each $B_{i,j}$ in terms of the entries of L. Be certain that you use a minimal number of operations.

6. Use the power iteration method to estimate the spectral radius for $(I - BA)$ for the two examples (Jacobi and Gauss-Seidel) developed in the text. How many iterations are required to complete the estimate given above?

7. The proof of Theorem 3.2.3 in the symmetric case. Given a symmetric A, there is an orthonormal basis of eigenvectors $\{v_1, ..., v_m\}$. Let λ_i denote the eigenvalue for v_i. Suppose that $|\lambda_1|$ is the spectral radius of A.

a. Prove that

$$\frac{A^k x_0}{\|A^k x_0\|} = x_{k+1}. \tag{2.3.1}$$

b. Write $x_0 = \alpha_1 v_1 + \sum_{i=2} \alpha_i v_i$. Prove that

$$A^k x_k = \lambda_1^k (\alpha_1 v_1 + \sum_{i=2} \frac{\lambda_i^k}{\lambda_1^k} \alpha_i v_i).$$

c. Prove that

$$\alpha_1 v_1 + \sum_{i=2} \frac{\lambda_i^k}{\lambda_1^k} \alpha_i v_i \to \alpha_1 v_1$$

as $k \to \infty$.

d. Prove that $x_{k+1} \to \beta v_1$. (Hint: for any real, $a/|a| = \pm 1$.)

8. Consider the Gauss-Seidel example developed in the section. Determine the dimension of the Krylov subspace.

2.4 Functions of Several Variables: Finding Roots and Extrema

The techniques we develop in this section are also referred to as Newton's method since they use derivatives and a single initial estimate to establish an iterative process to search for a root. These are properties shared with Newton's method introduced in Section 1.3.

As these procedures apply to differentiable functions $f : \mathbb{R}^n \to \mathbb{R}^m$, they apply to linear systems which are not square or to square linear systems whose coefficient matrix is singular. More generally, setting $g = f.f$, alternatively $g = f^2$ if $m = 1$, then the roots of f are minimums of g. Hence, we need only consider the problem of finding extrema in order to find roots.

The techniques developed in this section are applicable to optimal control theory and sensitivity analysis. Sensitivity analysis is of particular interest. Here you define a function f which measures an outcome from given independent (input) variables. However, the parameters necessary to express f may not be known with certainty. For instance a formula in finance may depend on the price volatility (the variance of a random variable). But it is often the case that the variance, σ^2, is not known exactly. Therefore we can add the parameters to the list of independent variables. Sensitivity analysis attempts to determine how the outcome will vary with changes in the input variables. In this way we determine the relative importance of separate items on the input side.

In another direction, these minimization techniques apply to web search technology. In that case, each person browsing the web has a penalty function. This function is determined by his/her prior tendencies. The browser returns the list of web pages that minimizes the penalty function. Of course, this max/min problem has perhaps a hundred thousand variables.

Do note that we have written this section for functions defined on $\mathbb{R}^2$. It is easy to extend these techniques to $\mathbb{R}^n$.

We begin by looking at an example. Consider $f(x,y) = x^2+y^2$ mapping $\mathbb{R}^2$ to $\mathbb{R}$. (See Figure 2.4.1). The graph of f is a subset of $\mathbb{R}^3$, and the single minimum of f is at $(0,0)$. Suppose we start the search for a minimum at $(1,2)$. If we think back to the method developed in Section 1.3. If we proceed analogously, we want a line γ tangent to the surface passing through $(1,2,5)$ where $5 = f(1,2)$. Then we want to determine where γ intersects the xy-plane. That point will be our next approximate minimum. If we repeat the process, we expect that we will get better and better approximate roots.

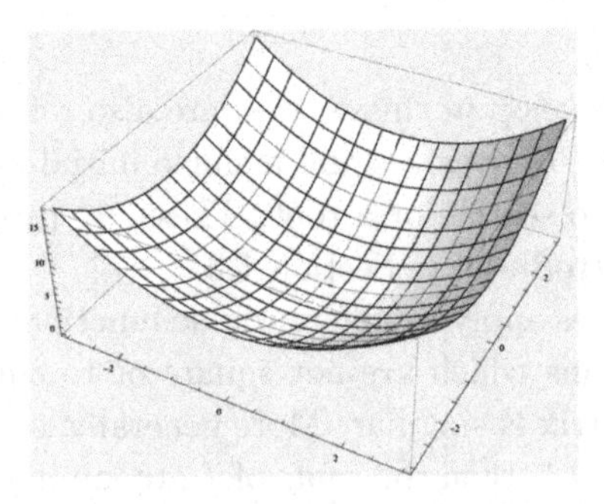

Figure 2.4.1: $f(x,y) = x^2 + y^2$

However, there are problems to overcome. The first problem is that we know the tangent plane at $(1,2,5)$. It has equation $z - 5 = \nabla f(1,2).(x - 1, y-2)$. But we do not know which line on the plane to use for γ. There are two standard procedures to determine the direction vector γ. We develop one now and the second at the end of the section. Recall a fact from multivariate calculus, the gradient points in the direction of maximal descent.

Hence, it seems reasonable to select that direction for the line γ. In this case, the technique is often called the method of *maximal descent.*

We know that the gradient $\nabla f = \partial f/\partial x(x_0, y_0)$ is a vector in the xy-plane, the domain of f, that determines the direction of maximal change for f. So, it is reasonable to set $\xi = \partial f/\partial x(x_0, y_0)$ and $\eta = \partial f/\partial y(x_0, y_0)$ and consider the line $(1, 2) + t(\xi, \eta) = (1 + t\xi, 2 + t\eta)$ in the xy-plane. Next, we define a function $h : \mathbb{R} \to \mathbb{R}$, $h(t) = f(1+t\xi, 2+t\eta)$. We can now solve for a max/min of h. This is a one variable calculus problem. Finding a minimum for h should yield a value for f less than 5, the value at $(1, 2)$. Indeed, $\nabla f(1, 2) = (2, 4)$, $h(t) = (1+2t)^2+(2+4t)^2 = 5+20t+20t^2$. The derivative of h is $20 + 40t$. It has critical at $t = -0.5$. Now $h(-0.5) = f(0, 0) = 0 < 5$. Indeed, we recognize the origin as the minimum of f. And we have arrived in one step.

We now state the general process for functions of several variables. Suppose we seek a minimum of f mapping $\mathbb{R}^n$ to $\mathbb{R}$.

(1) Compute the gradient of f at (x_0, y_0), $\nabla f(x_0, y_0) = (\xi, \eta)$.
(2) Set $h(t) = f(x_0 + t\xi, y_0 + t\eta)$.
(3) Solve the single variable calculus problem for h to yield t_0.
(4) Set $(x_1, y_1) = (x_0 + t_0\xi, y_0 + t_0\eta)$.
(5) If $f(x_0, y_0) < f(x_1, y_1)$, then exit (the process has failed).
(6) If the iteration count exceeds the maximum, exit (the process has failed).
(7) If $|f(x_1, y_1) - f(x_0, y_0)|$ is sufficiently small, exit (possible success).
(8) Go back to Step 1 using (x_1, y_1) as the seed.

It is interesting to note that maximal descent is sufficient for the minimization problems that occur in big data and machine learning applications.

To introduce the second technique, we look at another example. Suppose $f(x, y) = \cos^2(x)e^y + 1$. The minima for the function occur at the odd multiples of $x = \pi/2$. (See Figure 2.4.2). If we start the search at $(0, 1)$, then $t_0 = -35.7979$, $(x_0, y_0) = (0, -96.2817)$ and we are way out on the negative y-axis. Even though the value of f is nearly zero (about 1.4×10^{-47}, no further processing will take us any closer to an actual minimum. Hence, maximal descent has failed for this case.

We must consider an alternative choice for the search direction, γ. One choice is similar to the secant method. In this case, we begin with the

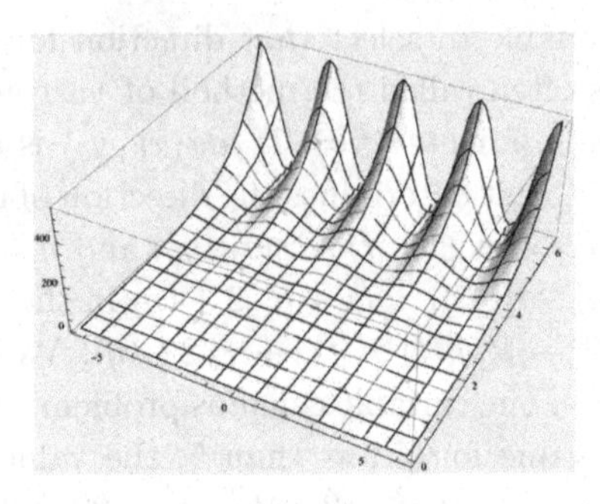

Figure 2.4.2: $f(x, y) = \cos^2(x)e^y$

Taylor expansion for f.

$$f(x+s) = f(x) + \nabla f(x).s + \frac{1}{2}s^T H(x)s + R_2 \tag{2.4.1}$$

where s^T denotes the transpose of s, H is the Hessian of f and R_2 is the the remainder term. Recall that the Hessian is the matrix whose entries are $\partial^2 f/\partial x_i \partial x_j$. Because of the use of the Hessian, this technique is referred to as the *Hessian method.*

If we suppose that $f(s) = f(x+s)$, then according to Rolle's theorem, we would expect a local extrema between x and $x+s$. Hence, $\gamma = s$ is the search direction. If we take the remainder term to be zero and we recast (2.4.1) for this case,

$$\frac{1}{2}H(x)s = -\nabla f(x). \tag{2.4.2}$$

Therefore, we solve for s. Since (2.4.2) is a linear system with coefficient matrix $H(x)$, then we can find γ provided $H(x)$ is nonsingular. We can now restate our procedure and describe the Hessian method. For this purpose, we need only replace statement 1 with the following

(1) Compute (ξ, η) as the solution to the linear system $0.5H(x)s = -\nabla f(x)$.

As mentioned at the beginning of the section, if f takes values in $\mathbb{R}^m$ then $g = f.f$ is real valued and the roots of f are now extrema for g. Hence, we can use the techniques developed here to solve the general problem $f(x) = 0$. We present examples in the exercises.

Exercises:

1. Use maximal descent to find a minimum for $f(x, y) = x^2 + xy + y^2$. Use $(2, 1)$ as the search starting point.

2. Use the Hessian method to find a minimum for $f(x, y) = x^2+xy+y^2$. Use $(2, 1)$ as the search starting point.

3. Let $f(x, y) = (x + y, x + y)$. Solve $f = 0$ using $(2, -1)$ as the initial estimate. Note that f is a singular linear transformation. When solving this problem you are solving a linear system with a singular coefficient matrix.

4. Consider the linear transformation

$$L(x, y, z, w) = \begin{pmatrix} 4 & -2 & 3 & -5 \\ 3 & 3 & 3 & -8 \\ -6 & -1 & 4 & 3 \\ -4 & 2 & -3 & 5 \end{pmatrix} \cdot \begin{pmatrix} x \\ y \\ z \\ w \end{pmatrix} = \begin{pmatrix} 4x - 2y + 3z - 5w \\ 3x + 3y + 5z - 8w \\ -6x - y + 4z + 3w \\ -4x + 2y - 3z + 5w \end{pmatrix}$$

a. Use LUDecomposition to determine if A is singular or non-singular. (Do not forget to introduce a decimal point to the data.) How does this impact the problem of solving an equation of the form $L(x, y, z, w) = (x_0, y_0, z_0, w_0)$?

b. Use the maximal descent method to solve $L(x, y, z, w) = (1, 1, 1, -1)$.

- Use $(5, 5, 5, 5)$ for the initial estimate.
- Use at least 35 iterations.
- Use 10^{-5} as the tolerance in Step 7.
- Make certain to use two 'if' statements, one for Step 5 and one for Step 7.

c. Redo Part b using $(1, 2, 3, 4)$ as the initial estimate.

d. Why is it possible for the solution to b and c to be different?

e. Prove that if v is the solution to b and $\hat{v}$ is the solution to c, then $v - \hat{v}$ solves $L(x, y, z, w) = (0, 0, 0, 0)$. (What is the kernel of a linear transformation?)

f. Use LinearSolve to get a solution to $L(x, y, z, w) = (1, 1, 1, -1)$. Is this solution trusted? Why? What was the condition number from Part a?

Chapter 3

Interpolating and Fitting

Introduction

We introduce the following terminology. Suppose we are given a set of n points $P_1, ..., P_n$ in the plane, $\mathbb{R}^2$, we may want to find a curve (function) which passes through the points (*interpolating*) or a curve which passes near to the points (*fitting*). If we want the curve to pass through the points, then we may have to accept anomalies on the curve. If we are willing to accept a curve that only passes near the points, then we may place stronger restrictions on the curve. In this chapter we see how this *give and take* materializes.

The several techniques include approximation via Taylor polynomials, polynomial interpolation, Bezier interpolation, spline interpolation, B-spline fitting, Hermite interpolation and least squares fitting. In most cases we include the error estimates.

Of the several techniques there is *no best of all*, no method that gives best results under all circumstances. The spline, with applications in computer graphics, visualization, robotics and statistics, is perhaps the most widely used. The spline curve is twice continuously differentiable, depends only on point data and faithfully reflects the tendencies of the input data. On the other hand, among the techniques we present, splines have the most complex mathematical foundation. For all of these reasons, we include the complete mathematical development of cubic splines.

In another direction, polynomial interpolation is the oldest of the techniques. It has the most developed theory and is widely used as a technique for approximating integrals and solutions to differential equations. For this reason, it is arguably the most important.

Least squares fitting in the linear case provides the numerical technique used for linear regression. Furthermore, least squares fitting often arises in the literature as a generalization of polynomial interpolation. In this context, it provides a technique for estimating the error for finite element method.

Another technique is Bezier interpolation. This procedure was developed originally to be used by engineers when resolving artist designs. In particular, Bezier curves were developed as a tool to help an engineer derive three dimensional coordinates from a designer's concept drawing.

The final technique is Hermite interpolation. In this case we are charged with finding a polynomial interpolation that both approximates the function but also its derivative. The Hermite interpolation provides the underlying mathematical foundation for Gaussian quadrature.

Before proceeding, we mention the theorem of Weierstrass, any continuous function on a closed interval is the uniform limit of a sequence of polynomial functions. This is a remarkable result. Also it is very old (originally proved in 1885). The proof however does not explain what the polynomials are. It was not until 1912 that Bernstein identified a sequence of polynomials [Rudin (1976)]. Even though the *Berstein polynomials* are determined by the continuous function, they do not interpolate the target in the sense that the function and the polynomial do not share designated points. Further, the convergence is very slow. A reasonable approximation of a function with Bernstein polynomials often requires Bernstein polynomials of degree two or three thousand. This is not a useful alternative to the techniques we are about to develop.

3.1 Polynomial Interpolation

In this section we introduce the idea of a polynomial interpolation. Given an unknown function f with known values at points x_i, we construct an interpolating polynomial p that agrees with f at these locations. The idea is, if p agrees with f at designated locations then we can use p in place of f. But if f is not continuous at x_i, then the values of f at the location may not relate to values of f at nearby points. Therefore, in order to talk about interpolation, we must first assume continuity. We make this assumption throughout.

We begin by looking at the Taylor expansion of a function. Consider the function $f(x) = xe^{-x} - 1$. Plotting this function on the interval $[1, 4]$:

```
f[x_] = x*Exp[-x] - 1;
```

```
Plot[f[x], {x,1,4}];
```

shows a decreasing function with an inflection point.

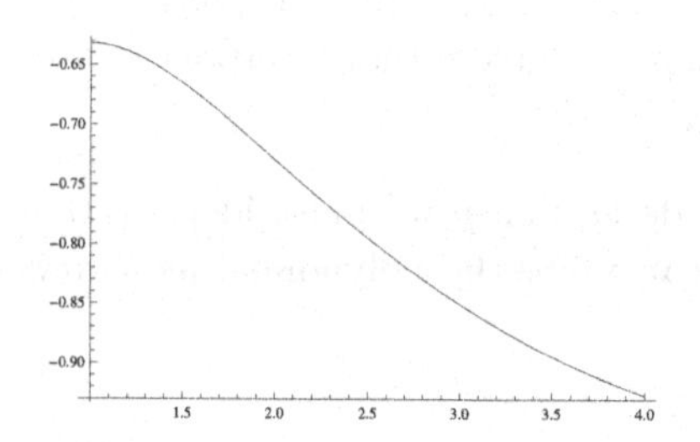

Figure 3.1.1: f together with the Taylor expansion at $x = 2.5$

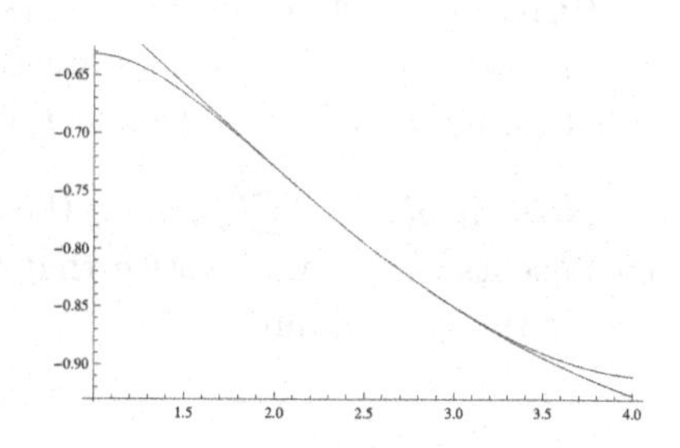

Figure 3.1.2: f concave down near the root

Thinking of this curve as being more or less cubic, we can develop the *cubic Taylor polynomial interpolation*for f expanded at the midpoint, $x = 2.5$.

$$g(x) = f(2.5) + \frac{df}{dx}(2.5)(x-2.5) + \frac{1}{2!}\frac{d^2f}{dx^2}(2.5)(x-2.5)^2$$
$$+\frac{1}{3!}\frac{d^3f}{dx^3}(2.5)(x-2.5)^3.$$

When developing g you will need to compute the derivatives of f. Recall that the derivatives of f are computed in *Mathematica* via $D[f[x], x], D[f[x], x, x]$ and so forth. If you plot f and g on the same axis you will see that the cubic Taylor polynomial provides a remarkably good approximation of this function. Figure 3.1.2 shows the graph of g together with the graph of f. Notice also that the graph of g is above f on the left and below on the right.

A numerical measurement of the *goodness of fit* is given by the L^2 norm of $f - g$,

$$\|f - g\|_2 = \Big(\int_1^4 (f-g)^2 dx\Big)^{1/2}.$$

This is called the *norm interpolation error.* In turn, the *mean norm interpolation error* is

$$\frac{1}{4-1}\Big(\int_1^4 (f-g)^2 dx\Big)^{1/2}.$$

The finite Taylor expansion produces a high quality one point interpolation provided we know the original function. However, suppose we have points and no function, then we will need a different approach.

Definition 3.1.1. Consider points $P_0, ..., P_n$ in $\mathbb{R}^2$, $P_i = (x_i, y_i)$. The *polynomial interpolation* is a polynomial p of degree n that interpolates the $n+1$ points in the sense that $p(x_i) = y_i$.

Now, if $p(x) = \sum_{i=0}^{n} \alpha_i x^i$, then to determine p we must identify the coefficients $\alpha_0, ..., \alpha_n$. Notice that we can write the polynomial as a row vector times a column vector,

$$(x^0, x^1, ..., x^n)\begin{pmatrix}\alpha_0\\ \alpha_1\\ ...\\ \alpha_n\end{pmatrix} = \sum_{i=0}^{n} \alpha_i x^i.$$

Our requirement for p is that it interpolates the $n+1$ points. Hence, we have for each i,

$$p(x_i) = (x_i^0, x_i^1, ..., x^n)\begin{pmatrix}\alpha_0\\ \alpha_1\\ ...\\ \alpha_n\end{pmatrix} = y_i$$

Collecting these equations we then get the following matrix equation

$$\begin{pmatrix}1 & x_1^1 & ... & x_1^n\\ . & . & ... & .\\ 1 & x_{n+1}^1 & ... & x_{n+1}^n\end{pmatrix}\begin{pmatrix}\alpha_0\\ \alpha_1\\ ...\\ \alpha_n\end{pmatrix} = \begin{pmatrix}y_0\\ y_1\\ ...\\ y_n\end{pmatrix}$$

where $x_i^0 = 1$. This is a linear system of equations where the x_i and y_i are known while the α_i are unknown. Hence, we can use the LinearSolve function in *Mathematica* to find the coefficients of p provided the coefficient matrix is non-singular. The matrix is called a *Vandermonde matrix*, it is known to be nonsingular provided the x_i are distinct.

Theorem 3.1.1. *The Vandermonde matrix*

$$\begin{pmatrix}1 & x_1^1 & ... & x_1^n\\ . & . & ... & .\\ 1 & x_{n+1}^1 & ... & x_{n+1}^n\end{pmatrix}$$

is nonsingular, provided the scalars x_i, $i = 0, 1, ..., n$ *are distinct.*

Proof. The Vandermonde matrix is singular only if the columns are dependent. In particular, only if there are scalars $\beta_0, ..., \beta_n$ not all zero with

$$\beta_0 \begin{pmatrix} 1 \\ ... \\ 1 \end{pmatrix} + \beta_1 \begin{pmatrix} x_1^1 \\ ... \\ x_{n+1}^1 \end{pmatrix} + ... + \beta_n \begin{pmatrix} x_1^n \\ ... \\ x_{n+1}^n \end{pmatrix} = \begin{pmatrix} 0 \\ ... \\ 0 \end{pmatrix}$$

Hence, for each $i = 1, 2, ..., n+1$ we have

$$\beta_0 + \beta_1 x_i^1 + ... + \beta_n x_i^n = 0.$$

In particular, we have demonstrated a polynomial $\beta_0 + \beta_1 x^1 + ... + \beta_n x^n$ that is not zero, has degree n and therefore has at most n distinct roots. However, we just showed that it has $n+1$ distinct roots, $x_1, ..., x_{n+1}$. As this is impossible, we are led to the conclusion that the Vandermonde matrix is nonsingular. □

There is another way to do polynomial interpolation. The outcome is the same, but nevertheless, the approach does provide insight. As in the previous case we begin with $n+1$ points in $\mathbb{R}^2$, denoted $P_1, ..., P_{n+1}$ with $P_i = (x_i, y_i)$. For each i, we set

$$l_i(x) = \frac{(x - x_1)(x - x_2)...(x - x_{i-1})(x - x_{i+1})...(x - x_{n+1})}{(x_i - x_1)(x_i - x_2)...(x_i - x_{i-1})(x_i - x_{i+1})...(x_i - x_{n+1})} = \frac{\prod_{j \neq i}(x - x_j)}{\prod_{j \neq i}(x_i - x_j)}.$$

It is not difficult to see that the polynomials $l_i(x)$ have degree n, satisfy $l_i(x_i) = 1$ and $l_i(x_j) = 0$ whenever $j \neq i$. Moreover, $q(x) = \sum_{i=1}^{n_1} y_i l_i(x)$ interpolates the given points. (See Exercise 5.) The polynomials $l_i(x)$ are called *Lagrange polynomials.* We now see that the two polynomial interpolations, p derived from the Vandermonde matrix and q derived from the Lagrange polynomials are in fact the same.

Theorem 3.1.2. *We are given interpolation points $P_1, ..., P_{n_1}$ associated to a continuous function f. Suppose that the interpolation derived from the Vandermonde matrix is given by p and the interpolation derived from the Lagrange polynomials by q, then $p(x) = q(x)$.*

Proof. We begin by setting $r = p - q$. Hence, r is a degree n polynomial. Since $p(x_i) = y_i = q(x_i)$ for each $i = 1, 2, ..., n+1$, then r has $n+1$ roots, $x_1, ..., x_{n+1}$. But, if r is not identically zero, then it can have at most n roots. Therefore, $r = 0$ and $p = q$. □

It is possible that you must use Lagrange polynomials to compute the polynomial interpolation of a function. In particular, there are cases where the Vandermonde matrix procedure does not work. Suppose that two of the x-axis locations x_i and x_j are very close together. Then it would appear to *Mathematica* that the corresponding rows of the Vandermonde matrix are equal or nearly equal. In this case the condition number will be large and LinearSolve will not return reliable results. Nevertheless, it is still possible to get the interpolation via Lagrange polynomials.

If the points P_i lie on the graph of a function f, then it is natural to ask how well does p approximate f. If f has at least $n+1$ continuous derivatives then we can estimate the error, $e(x) = f(x) - p(x)$. Recall that with this hypothesis, then the Taylor interpolation error will have a known value.

Theorem 3.1.3. *Suppose that f is a real valued function defined on an interval $[a,b]$ and suppose that f has at least $n+1$ continuous derivatives. Further, take $a \leq x_1 < \ldots < x_{n+1} \leq b$, with $f(x_i) = y_i$. If p is the polynomial interpolation of the points (x_i, y_i), then the error $e(x) = f(x) - p(x)$ is given by*

$$e(x) = \frac{f^{(n+1)}(\xi)}{(n+1)!} \prod_i (x - x_i), \tag{3.1.1}$$

for some $\xi = \xi_x$ in (a,b) depending on x. In particular,

$$|e(x)| \leq \frac{M}{(n+1)!}(b-a)^{n+1}, \tag{3.1.2}$$

where M is the maximal value of $f^{(n+1)}$ on the interval.

Proof. We define $g(x) = e(x)/\prod_i(x - x_i)$, so that $e(x) = f(x) - p(x) = \prod_i(x - x_i)g(x)$. Next, take ζ in $[a,b]$ distinct from the x_i and set

$$h(x) = f(x) - p(x) - \prod_i (x - x_i) g(\zeta).$$

Note that we cannot be certain that g is defined at the x_i, however our choice of ζ assures us that h is defined on $[a,b]$ with $n+1$ continuous derivatives.

Now, each x_i is a root of h and in addition $h(\zeta) = e(\zeta) - \prod_i(\zeta - x_i)g(\zeta) = 0$. Hence, h has $n+2$ roots in the interval $[a,b]$. Furthermore, h is continuous on the closed interval and differentiable on the open interval (a,b). Hence, we may apply Rolle's theorem to the interval between each pair of successive roots. In particular, we conclude that between each pair of roots there is a root of the derivative of h. Hence, dh/dx has at least $n+1$ roots on the interval (a, b). Repeating this argument, d^2h/dx^2 has at least n roots in (a, b). Continuing, the kth derivative of h has at least $n + 2 - k$ roots. So that the $n + 1^{st}$ derivative has at least 1 root. We denote this root by $\xi = \xi_\zeta$, since ξ depends on our choice of ζ. Now

$$0 = h^{(n+1)}(\xi) = f^{(n+1)}(\xi) - p^{(n+1)}(\xi) - g(\zeta)\frac{d^{n+1}}{dx^{n+1}}\prod_i(x - x_i)|_{x=\xi}.$$

But p is degree n, so $p^{(n+1)} = 0$. Also $d^{n+1}/dx^{n+1}\prod_i(x - x_i) = (n + 1)!$, no matter what ξ is. Therefore,

$$e(\zeta) = \frac{f^{(n+1)}(\xi)}{(n+1)!}\prod_i(\zeta - x_i).$$

Finally, since h is defined for any x in the interval, then this last expression for the error is satisfied for all x.

For the final statement on the bound for the error magnitude we note that since f is $n+1$ times continuously differentiable, then $f^{(n+1)}$ is continuous and hence has maximum value on the interval. □

Numerical integration is based on polynomial interpolation. Hence, the interpolation error is also the numerical integration error. In turn, polynomial interpolation is an important feature in approximating the solution of a partial differential equation. Hence, interpolation error arises in that context. On the other hand, the estimate for the error magnitude is of little use if we do not have information about f. Indeed, it is not difficult to find functions where M is very large. Nor is it difficult to find functions where the error is large. The following example is a case in point.

Returning to the function $f(x) = xe^{-x} - 1$ and the four points $P_i = (x_i, y_i)$, $x = 1, 2, 3$ and 4. The polynomial interpolation, $p(x)$, of the points will again provide an approximation of f by a cubic polynomial. As in the case of the Taylor interpolation, it is remarkably close to f. On the other hand consider the function $f(x) = 1/(1 + x^2)$. In this case, pick a finite sequence of points along the graph of f, which are symmetric about the y-axis. Use these points to produce a polynomial interpolation of the f. (See Exercise 3 below.) The problem is that the polynomial looks nothing like the function. Further, the more points you choose the less the polynomial resembles f. Looking at the graph if f we see that the function seems not to be a polynomial function. (Note the asymptotic behavior. It is not easy to find a polynomial that can reproduce this type of behavior.) Hence, we should not expect that there is a polynomial function that interpolates it well.

There is another problem with polynomial interpolation. Consider again the function $f(x) = 1/(1+x^2)$ and select four points $P_1 = (-4, 1/17), P_2 = (-2, 1/5), P_3 = (2, 1/5), P_4 = (4, 1/17)$ from the graph of f. Next select $P = (0, y)$ where $y \in [0.2, 0.3]$. Figure 3.3 shows the resulting polynomials for three values of y. Suppose that the location of the points came from some measuring or sampling process, then small errors (as in this case) may yield significantly different results. Looking at the resulting curves we see that shape of the curves is different. Further the change in y is magnified 20 times at $p(5)$. This is an inherent problem with polynomial interpolation. The technical term for the problem is that polynomial interpolation lacks *local control*. In a subsequent section we develop spline curves. These curves were developed precisely to resolve the local control problem.

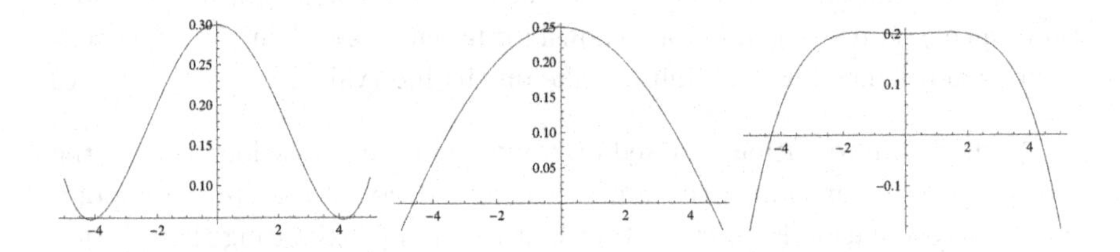

Figure 3.1.3: Three alternate images $y = 0.3,\ 0.25,\ 0.2$; $p(5) = 0.11,\ -0.04,\ -1.9$

In spite of the problem we just noted, polynomial interpolation is an important and productive tool for numerically solving differential equations. When this technique is used special care is taken to ameliorate the problem we see in Figure 3.1.3.

Because the Taylor expansion requires more information than is usually available, it is often ignored as an interpolation technique. However, there is an important application, which should not be ignored. In the next section we will develop a class of parametric cubic interpolations. Consider the setting where $\beta(t) = (\beta_1(t), \beta_2(t))$ in $\mathbb{R}^2$ and each β_i is an ordinary cubic polynomial. When β represents a function, then it is possible to solve $x = \beta_1(t)$ for t and then substitute this in β_2 to yield $\beta = (x, f(x))$. However, the resulting function is rarely integrable. On the other hand, you can get values for f and its derivatives. Hence, you can write the cubic Taylor expansion for f and this is easily integrated.

Finally, in Exercise 7 we introduce the idea of piecewise polynomial interpolation. The basic idea of polynomial interpolation is that the more points that we interpolate, then the better the polynomial will approximate the original function. However, as we add more and more points then the degree of the polynomial increases. In piecewise polynomial interpolation, we subdivide the interval into smaller and smaller subintervals while interpolating the function by polynomials of fixed degree on each subinterval. The later option is often preferred when simulating processes arising from a PDE.

Exercises:

1. Compute the norm error and the mean norm error for the function $f(x) = xe^{-x} - 1$ and its cubic Taylor expansion about $x = 2.5$. Use the interval $[1, 4]$.

2. For $f(x) = xe^{-x} - 1$.

a. Compute the polynomial interpolation p for the $P_iv = (x_i, y_i)$, $x = 1, 2, 3$ and 4.

b. Plot the graph of f and p on the same axes for the interval $[1, 4]$.

c. Compute the norm error and mean norm error.

d. Is p better or worse than the cubic Taylor interplant.

e. Use Equation 3.1.2 to estimate the maximal absolute error for the interpolation p.

3. Compute the polynomial interpolation of the points $(x_i, 1/(1 + x_i^2)$ for $x_i = -2, -1, 0, 1, 2$. Plot the polynomial against the graph of $f(x) = 1/(1 + x^2)$. Compute the norm error and mean norm error.

4. Repeat Exercise 3 with additional points on the x-axis, add 3, -3, 4 and -4. Does this produce a better approximation of the function f?

5. Consider the points $P_i = (x_i, y_i)$, $i = 1, 2, ..., n + 1$, in the real plane and the corresponding Lagrange polynomials l_i.

a. Prove that for each i, l_i is a degree n polynomial with $l_i(x_j) = 0$ if $i \neq j$, and $l_i(x_i) = 1$.

b. Prove that $q(x) = \sum_{i=1}^{n+1} y_1 L_i(x)$ interpolates the given points.

c. Prove that the Lagrange polynomials are linearly independent in the vector space $\mathbb{P}_n$ of polynomials of degree no larger than n.

6. Use Theorem 3.1.3 to estimate the maximal absolute error in Exercises 3 and in 4. Does adding the additional points increase or decrease the error estimate?

7. Consider the function $f(x) = 1/(1+x^2)$ on the interval $[-4, 4]$.

a. Determine the maximal value for $\mu = f^{(3)}(x)$ on the interval.

b. Divide the interval into 40 subintervals of length 0.2. In particular determine $-4 = a_0 < a_1 < ... < a_{40}$ with each $a_{k+1} = a_k + 0.2$.

c. Compute the second degree polynomial interpolation of f on the subinterval $[a_k, a_{k+1}]$ using the three values a_k, $(a_k + a_{k+1})/2$ and a_{k+1}.

d. Plot the result of Part c and overlay the plot of f.

e. Use Theorem 3.1.3 to prove that the absolute error $|e(x)|$ is bounded by $\mu/6(0.2)^{40}$, where μ is the value computed in Part a.

f. Prove that as the number of subintervals goes to ∞, then error converges to zero.

8. Researchers writing in a chemical engineering journal reported the following data on tungsten production as a function of temperature measured in degrees Kelvin.

t	700	800	900	1000
$f(t)$	0.071	0.084	0.097	0.111

They determined that the data fit to the following function (to 3 decimal places accuracy),

$$f(t) = 0.02424(t/303.16)^{1.27591}.$$

a. Use a cubic polynomial to interpolate the given data. Use this polynomial to estimate the values at $t = 750, 850$ and 950. (Because the values of t are large, the Vandermonde matrix will appear to Mathematica as ill conditioned. You resolve this problem by using Lagrange polynomials.)

b. Assuming that f is the correct predictor for tungsten, determine the mean absolute error for the three estimates in Part a.

c. Again assuming that f is correct, use Equation 3.1.2 to calculate the estimated error for the cubic polynomial interpolation as a function of t. Then determine the estimated mean absolute error for the three values of t.

d. Is the actual mean absolute error smaller than the estimated mean absolute error?

3.2 Bezier Interpolation

Bezier interpolation arose in 1962 to solve a problem in the manufacturing industry. When a new product is begun, a designer will produce a rendering. Engineers will then produce specifications from the designer's drawing. In the automobile or aircraft industries, the engineers task was indeed difficult. The designer would produce a concept drawing of the car or aircraft. From this drawing, the engineers would have to specify the requirements for the sheets of metal for the exterior, the necessary frame and then they could infer the shapes and sizes of the spaces for the passenger compartments, the engine compartment, etc. The task was nearly impossible causing cost overruns and time delays.

The tools at their disposal were primitive. Often they would produce a wooden model of the object, then slice the wooden object with a saw to create a sequence of cross sections. Next the cross sections where projected on a large screen at actual size, so that images could be traced and measured to yield the data necessary for construction.

This was the context when Bezier interpolation was introduced. Pierre Bezier was an engineer at Renault. He is credited with the process. However, it is generally assumed that Lockheed Aircraft already had the process but kept it secret. On the other hand, Citroen had a less capable but related process.

The Bezier curve is a parametric cubic curve based on 4 guide points B_1, B_2, B_3 and B_4. The curve has end points at B_1 and B_4, and tangent lines at these end points passing through B_2 and B_3. If we designate the parametric curve in $\mathbb{R}^2$ as $\beta(t) = (\beta_1(t), \beta_2(t))$ with t in $[0, 1]$, then these requirements may be stated as

$$\beta(0) = B_1, \beta(1) = B_4; \ \frac{d}{dt}\beta(0) = 3(B_2 - B_1), \frac{d}{dt}\beta(1) = 3(B_4 - B_3). \tag{3.2.1}$$

Bezier, proposed the following. It is based a geometric construction. First, fix the four points, B_1, B_2, B_3 and B_4, called *guide points*. Next, connect the four guide points with line segments. (See Figure 3.2.1a.) Fix a real t in the interval $[0, 1]$. Using t, locate the point $B_{1,1} = B_1 + t(B_2 - B_1)$ on the line segment connecting B_1 to B_2. Similarly, locate $B_{1,2}$ and $B_{1,3}$ between B_2, B_3 and B_3 and B_4, respectively, and connect these points with line segments. (See Figure 3.2.1b.) Repeat the process with the three points $B_{1,1}, B_{1,2}$ and $B_{1,3}$ to derive two additional points $B_{2,1}$ and $B_{2,2}$ on the segments connecting $B_{1,1}$ to $B_{1,2}$ and $B_{1,2}$ to $B_{1,3}$. Finally, we set

$\beta(t) = B_{2,1} + t(B_{2,2} - B_{2,1})$. (See Figure 3.2.1c.) If we write $\beta(t)$ in terms of the original four points, we have the usual representation for the *Bezier curve.*

$$\beta(t) = (1-t)^3 B_1 + 3t(1-t)^2 B_2 + 3t^2(1-t)B_3 + (t^3)B_4. \qquad (3.2.2)$$

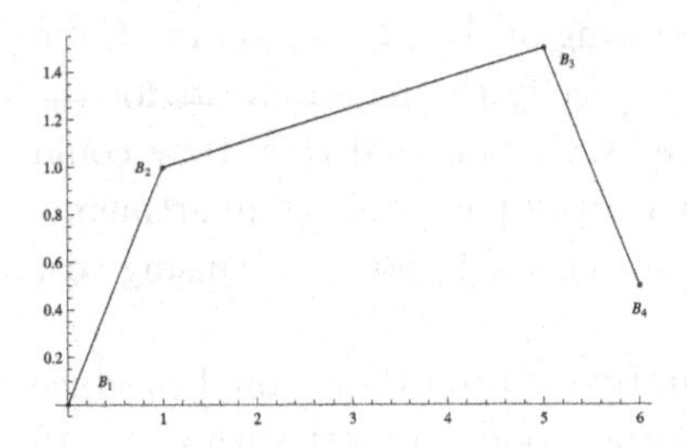

Figure 3.2.1a: Four guide points with segments

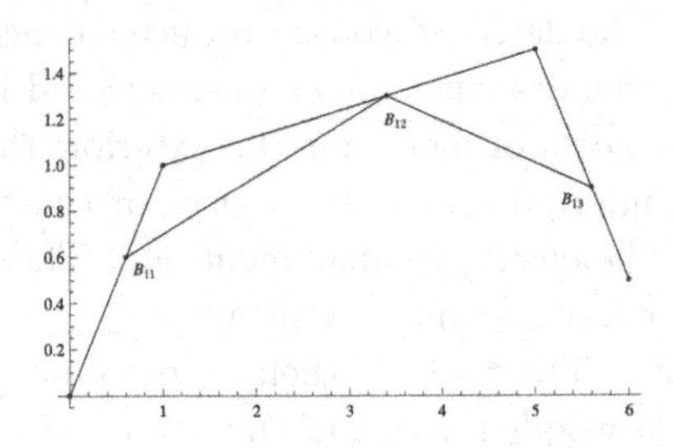

Figure 3.2.1b: 2nd-level points with line segments, t = 0.6

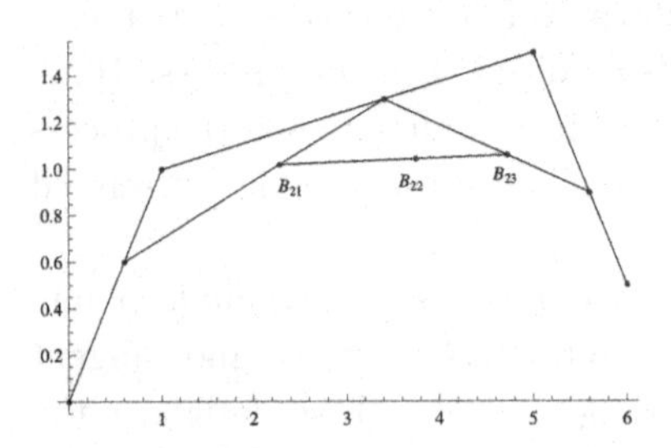

Figure 3.2.1c: $3^r d$-level points with line segments

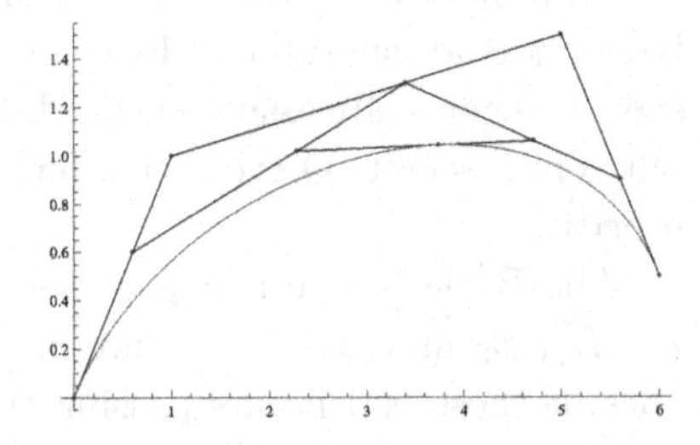

Figure 3.2.1d: The Bezier curve $\beta(t)$

Letting t vary in the unit interval, (3.2.2) describes a parametric cubic. Figure 3.2.1d shows the plot of $\beta(t)$. The associated coordinate functions are given by

$$\beta_1(t) = (1-t)^3 x_1 + 3t^2(1-t)x_2 + 3t(1-t)^2 x_3 + t^3 x_4,$$

$$\beta_2(t) = (1-t)^3 y_1 + 3t^2(1-t)y_2 + 3t(1-t)^2 y_3 + t^3 y_4,$$

where $B_i = (x_i, y_i)$.

Just as in the case of the polynomial interpolation, we require four points to do a cubic interpolation. However, in this case the necessary information includes only two points on the curve (the starting point and ending point) and information determining the slope of the curve at these two points. More complicated curves can be constructed by piecing successive Bezier curves together.

Return to the function $f(x) = xe^{-x} - 1$. We can derive the Bezier interpolation of f by setting $B_1 = (1, f(1))$ and $B_4 = (4, f(4))$ and using the derivative of f at 1 and 4 to determine the other two guide points. Since the points on the function graph are given by $(x, f(x))$, then the tangent vectors to the graph are $d/dx(x, f(x)) = (1, f'(x))$. Hence, the tangent vector at B_1 is $(1, f'(1))$. By (3.2.1), this vector must also satisfy $(1, f'(1)) = 3(B_4 - B_3)$. Therefore, $B_2 = B_1 + 1/3(1, f'(1))$. Similarly, by (3.2.1), $B_3 = B_4 - 1/3(1, f'(4))$.

As in the previous cases, the Bezier interpolation of f yields a good approximation of the original curve.

In Section 1.3 we showed a curve for which Newton's method failed because the process cycled, the third estimated root was equal to the first, the fourth equal to second and so forth. We created this curve using a Bezier curve. We started with $B_1 = (-1, 1)$ and $B_4 = (1, 1)$. Next, we wanted the slope at B_1 to be $-1/2$ and equal to $1/2$ at B_4, so that Newton's method would return points $(1, 0)$ and $(-1, 0)$. Using the technique described above, we have

$$(-1, -1/2) = 3(B_2 - B_1); (1, 1/2) = 3(B_4 - B_3).$$

One purpose of interpolating points is to use the interpolating function to compute the integral of a unknown function inferred from the points. We will see later, that the parametric form of the Bezier curve is significantly more difficult to deal with than the polynomial or Taylor interpolation.

Exercises:

1. Use the Bezier technique to interpolate $f(x) = xe^{-x} - 1$ for $x \in [1, 4]$. Plot f and β on the same axes.

2. Interpolate $f(x) = 1/(x^2 + 1)$ between -2 and 2. Use one Bezier curve between -2 and 0 and another between 0 and 2.

3. Do the illustration at the end of Section 1.3. If the curve is given by $\beta(t) = (\beta_1(t), \beta_2(t))$, then use the *Mathematica* statement, $Solve[\beta_1(t) == x, t]$, to solve for x in terms of t. Insert the result into $\beta_2(t)$. The result will give you an expression for the curve in the form $f(x) = y$. Use *Expand* to fully resolve f. Plot f.

4. In the Introduction to Chapter 3, we mentioned the Bernstein Polynomials. Given a set of points $P_0, P_1, ..., P_n$ with $P_i = (x_i, y_i)$ in the plane, then the nth Bernstein polynomial is given by

$$p_n(t) = \sum_{i=0}^{n} \frac{n!}{i!(n-i)!}(1-t)^i t^{n-i} P_i.$$

a. Prove that the cubic Bezier curve $\beta(t)$ defined on four points is identical with the third Bernstein polynomial.

b. Prove that for any $n, p_n(0) = P_0$ and $p_n(1) = P_n$.

c. Use parts a, b to define a generalization of cubic Bezier curves.

3.3 Least Squares Fitting

We begin with a word of caution. Least square fitting in the linear case arises also in the context of linear regression. This is more of a coincidence than anything else. It is true that in both cases a line is fit to a finite set of points. In addition, the line arises from the same minimization process. Beyond that the processes are different and distinct. Least squares fitting in the numerical methods context is a procedure that begins with a set of points, and then guides the researcher to a polynomial, which seems to fit well to the points. We will see that it may be considered as a generalization of polynomial interpolation. On the other hand linear regression begins with a set of points sampled from a distribution and includes assumptions on the distribution and the sample. Then a line is inferred. In particular, the line is derived by minimizing the variance of a related distribution. Furthermore, statistics are returned indicating confidence intervals for the slope and y-intercept of the line, as well as, a general statistic indicating whether linear regression was a reasonable approach to the data.

In short, least squares fitting is a process that begins with a set of points and returns a best fitting polynomial. Regression is a statistical process that applies to a sample from a distribution, fits a line to the sample and returns statistical information about the reliability of the process. In this course we are concerned only with the least squares process.

We begin with points $P_1, ..., P_n$ with $P_i = (x_i, y_i)$ for each i. We are expecting to find a line, $y = mx + b$, which best fits the point set. In order to proceed we must define *best fits*. Indeed, the term least squares refers to the following definition of *best fit*. Suppose we were to calculate the vertical distance from each of the points to the line and then total the squares of all the distances. We will say that a line best fit the point set, if this number (sum of squared vertical displacements) is minimal. Notice that we have described a calculus max/min problem.

With this description in mind, we write out the term for the total calculated displacement as a function of the slope and y-intercept of the line. First, the vertical distance from $P_i = (x_i, y_i)$ to the line $y = mx + b$, is $|y_i - (mx_i + b)|$. Since we are heading toward a calculus style max/min process, the absolute value is inconvenient. Hence, square each of these terms and then add to get a function σ with independent variables m and b.

$$\sigma(m, b) = \sum_i (y_i - mx_i + b)^2.$$

The next step is to apply standard max/min techniques to σ. To begin, we differentiate the dependent variable with respect to the independent variable.

$$\frac{\partial \sigma}{\partial m} = \sum_i 2(y_i - (mx_i + b))(-x_i) = 2(\sum_i (mx_i^2 + (b - y_i)x_i),$$

$$\frac{\partial \sigma}{\partial b} = \sum_i 2(y_i - (mx_i + b))(-x_i) = 2\sum_i (y_i - (mx_i + b)).$$

Setting these two terms to zero and reorganizing them just a little, we get

$$0 = \sum_i (mx_i^2 + (b - y_i)x_i) = m\sum_i x_i^2 + b\sum_i x_i - \sum_i x_i y_i$$

$$0 = \sum_i (y_i - (mx_i + b)) = -m\sum_i x_i - nb + \sum_i y_i.$$

Now, these two equations can be recast as a 2 by 2 linear system with unknowns m and b.

$$\sum_i x_i y_i = (\sum_i x_i^2)m + (\sum_i x_i)b$$

$$\sum_i y_i = (\sum_i x_i)m + nb.$$

Or in matrix notation,

$$\begin{pmatrix} \sum_i x_i^2 & \sum_i x_i \\ \sum_i x_i & n \end{pmatrix} \begin{pmatrix} m \\ b \end{pmatrix} = \begin{pmatrix} \sum_i x_i y_i \\ \sum_i y_i \end{pmatrix}.$$

Next, set

$$A = \begin{pmatrix} x_1 \dots x_n \\ 1 \dots 1 \end{pmatrix}.$$

Then it is immediate that

$$AA^T = \begin{pmatrix} x_1 \dots x_n \\ 1 \dots 1 \end{pmatrix} \begin{pmatrix} x_1 & 1 \\ \cdot & \cdot \\ x_n & 1 \end{pmatrix} = \begin{pmatrix} \sum_i x_i^2 & \sum_i x_i \\ \sum_i x_i & n \end{pmatrix},$$

and

$$A \begin{pmatrix} y_1 \\ \cdot \\ y_n \end{pmatrix} = \begin{pmatrix} \sum_i x_i y_i \\ \sum_i y_i \end{pmatrix}.$$

Hence, we may rewrite the 2 by 2 system as

$$AA^T \begin{pmatrix} m \\ b \end{pmatrix} = A \begin{pmatrix} y_1 \\ \cdot \\ y_n \end{pmatrix}.$$

This form of the linear system is most suitable for our calculations. It is straight forward to prove that the coefficient matrix, $A(A^T)$, is necessarily nonsingular provided that no two x_i are equal. See Exercise 2.

To get a feel for how this looks, consider the following example. Suppose we have points $P_1 = (-5, 3), P_2 = (-4, 2), P_3 = (-2, 7), P_4 = (0, 0), P_5 = (1, 5), P_6 = (3, 3), P_7 = (5, 5)$. The following figures shows the points and the resulting least squares line.

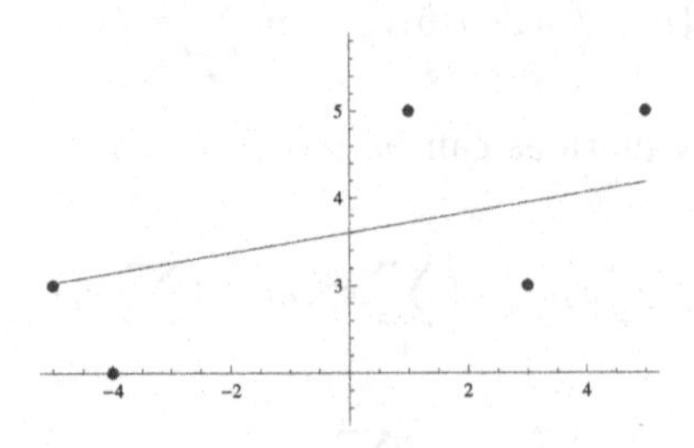

Figure 3.3.1: Even data points with least square fit

If we had asked for a quadratic polynomial that best fit the point set, then we would be looking for three coefficients, a, b and c. Setting up the problem as in the linear case to get $\sigma(a, b, c)$, differentiating with respect to the three variables, setting the resulting terms to zero and solving we would get the following linear system.

$$AA^T \begin{pmatrix} a \\ b \\ c \end{pmatrix} = A \begin{pmatrix} y_1 \\ . \\ y_n \end{pmatrix},$$

where

$$A = \begin{pmatrix} x_1^2 \ ... \ x_n^2 \\ x_1 \ ... \ x_n \\ 1 \ ... \ 1 \end{pmatrix}.$$

There are similar expressions for the cubic least squares problem, etc. The data shown in Figure 3.3.1 would seem to be cubic in Exercise 1 we investigate several possibilities. The result suggests that cubic is best.

At the top of the section we mentioned that least squares fitting was separate and distinct from linear regression. Before ending the section we add some details to that statement. The setting for linear regression starts with two random variables, X and Y, together with the hypothesis that Y is a linear function of X. In particular, we are proposing that $Y = aX + b$, where the parameters a and b are to be determined. Then the process is to select the parameters so as to minimize the variance of $Y - X$. When you do this calculation against sample data (supposing that the sample was done with replacement), the process is exactly the degree 1 least squares fitting. However, within the statistical context, the process returns values that measure the correctness of the hypothesis and provide confidence intervals for the two parameters. These are ideas special to statistical regression not shared with numerical analysis.

In the numerical analysis context there is no means to measure the correctness of the fit and no confidence intervals for the parameters. However, least squares fitting is used to approximate the solution to a partial differential equation. In this case, the points that drive the least squares fitting arise from numerical processes. There is a means to measure how well these values approximate the actual values and then use the least squares process to fill between the known data.

We end this section with an important application. Exponential growth is common in biology as well as the other sciences. For instance bacterial growth is exponential. Epidemics show exponential growth during their early stages. Exponential growth is characterized by the statement that the rate of change of population size is proportional to the current size. In particular, if $f(t)$ represents the number of organisms in a bacteria growth at time t, then *rate of change for f is proportional to the value of f* means that $df/dt = \gamma f(t)$. Hence, by integrating both sides, we get $f(t) = \alpha \int_0^t f(s)ds$ or $f(t) = \alpha e^{\beta t}$, where $\gamma = \alpha\beta$ and $\alpha = f(0)$.

Next, we turn this situation upside down. Suppose we have pairs (t_i, y_i) of data, which because of the setting we know to be related via an exponential, $y_i = \alpha e^{\beta t_i}$, but we do not know α and β. We can solve this problem with least squares fitting. We write $y = \alpha^{\beta t}$ and take the log of both sides. This yields, $\log[y] = \log[\alpha] + \beta t$. In this form, $\log[y]$ is a linear function of t. Hence, we have the technique. First, we take the log of the y_i, then fit these values to the t_i using a linear least squares fitting. The result is a line, $y = at + b$. Finally, we set $b = \beta$ and $\alpha = e^a$. Exercise 6 is an example of this sort of problem.

Exercises:

1. Fit the data $P_1 = (-5,3), P_2 = (-4,2), P_3 = (-2,7), P_4 = (0,0), P_5 = (1,5), P_6 = (3,3), P_7 = (5,5)$ to a line (Figure 3.3.1), a quadratic and a cubic. In each case, calculate the sum of squares from the curve to the points. Which curve gives the best fit.

2. For the linear case prove that AA^T is non-singular provided that there is a pair i and j, with $x_i \neq x_j$. The proof proceeds through the following steps.

a. Let $(x_1, ..., x_n)$ and $(y_1, ..., y_n)$ be elements of $\mathbb{R}^n$. Prove that $(\sum_i x_i y_i)^2 = (\sum_i x_i^2)(\sum_i y_i)^2 - \sum_{i<j}(x_i y_j - x_j y_i)^2$.

b. Set each $y_i = 1$ in Part a and conclude that $(\sum_i x_i)^2 = n\sum_i x_i^2 - \sum_{i<j}(x_i - x_j)^2$.

c. Prove that AA^T is singular if and only if $(\sum_i x_i)^2 = n\sum_i x_i^2$.

d. Conclude that AA^T is nonsingular provided there is a pair i and j, with $x_i \neq x_j$.

3. State and prove a result analogous to Exercise 2 for the case of the quadratic least squares fitting.

4. Suppose that $P_1, ..., P_n$ are points in $\mathbb{R}^3$, $P_i = (x_i, y_i, z_i)$ and let that $f(x, y) = a + bx + cy + dxy$. Derive a procedure to determine a, b, c and d so that $\sum_i \|(x_i, y_i, f(x_i, y_i)) - (x_i, y_i z_i)\|^2$ is minimized. Note that we use $\|(x, y, z)\|$ to denote the usual Euclidean length $(x^2 + y^2 + z^2)^{1/2}$.

5. Prove for the least squares fit of $n+1$ points to an nth degree polynomial A is the transpose of the Vandermonde matrix. Conclude the least squares fitting for this case is equivalent to the polynomial interpolation process described in Section 3.1.

6. The following data is known to be related via an exponential, $y = \alpha e^{\beta t}$. Use the procedure described in the section to identify α and β.

t	13.3522	10.6354	14.1693	9.67667	16.0862
y	10.8308	14.5888	31.5388	47.2135	53.374

t	6.221	4.93898	11.4164	9.13563	13.7273
y	1.66611	2.41672	4.70308	4.62067	7.95687

t	18.2355	13.243	16.2672	20.2108	15.1733
y	67.8773	98.8846	155.689	274.55	494.796

t	22.8416	23.3023	21.7688	23.9842	25.4667
y	754.759	758.462	1540.63	2059.3	2380.1

Plot the data along with $y = \alpha e^{\beta t}$ on the same axis.

7. This exercise is an extension of Exercise 8, Section 1. Recall that we had a function

$$f(t) = 0.02424\left(\frac{t}{303.16}\right)^{1.27591}$$

that predicted the amount of tungsten production as a function of temperature. Temperature is measured in degrees Kelvin. The following data is an extension of the data given before.

a. Plot the data. By inspection determine the degree of least squares that would fit the data best (degree 1, 2, or 3). Explain your decision.

b. Execute the least squares fitting and use it to predict the three values, $f(750)$, $f(850)$ and $f(950)$.

c. Determine the mean absolute error and compare the result to the previous result. Which method was better?

t	300	400	500	600	700	800
$f(t)$	0.024	0.035	0.046	0.058	0.071	0.084

t	900	1000	1100	1200	1300	1400
$f(t)$	0.097	0.111	0.126	0.14	0.155	0.17

t	1500	1600	1700	1800	1900	2000
$f(t)$	0.186	0.202	0.219	0.235	0.252	0.269

3.4 Cubic Splines and B-Splines

The term spline refers to a large class of curves. Some interpolate the given points while others fit the data. They are commonly used in many areas of application, including statistics, probability, engineering and computer graphics. One reason for their wide use is that splines exhibit local control. Another advantage to cubic spline interpolation or fitting over polynomial interpolation is that the curve can simulate asymptotic behavior. For this reason these curves are often used in probability theory to approximate a density function. Perhaps the most important advantage is smoothness. Splines are C^2; whereas, piecewise Bezier curves are only C^0.

In this section we develop two classes of spline, the classical cubic spline [Su and Liu (1989)] and the B-spline. In addition, cubic Hermite interpolation introduced in conjunction to Gaussian quadrature is often referred to as cubic orthogonal spline interpolation. We see this interpolation method in the next section. The Bezier curve is a parametric curve. However, the spline is formed from several segments where each segment is a parametric cubic. We state precisely,

Definition 3.4.1. Consider $[a,b] \subset \mathbb{R}$ with a partition $a = t_0 < t_1 < ... < t_n = b$. A *spline* σ defined on $[a,b]$ is a paramedic curve taking values in $\mathbb{R}^m$ such that

a. for each $i = 1, ..., n-1$, σ restricted to (t_{i-1}, t_i) is a parametric cubic, denoted σ_i,

b. $\sigma_i(t_i) = \sigma_{i+1}(t_i)$, $i = 1, ..., n-2$,

c. σ is twice differentiable at each t_i, $i = 1, 2, ..., n-2$.

The points $\sigma(t_i)$ are called the *knot points* and the σ_i are called the *segments*.

The splines developed in this section will be twice continuously differentiable at the knot points. Before continuing, we remark that Bezier curves are splines with a single segment.

The original use of the term spline arose in drafting. A spline was a thin wooden strip that a draftsman bends and pins on his table to provide a firm edge against which he could draw a curve. As we shall see, the parametric cubic is derived from the physical properties of the wooden object. The result is a linear system with size equal to the number of interpolation points.

We begin with the mathematical analysis of the draftsman's spline. Specifically, this spline is a thin elastic strip (traditionally made of wood), which is attached to a drafting table by pins and bent by weights (called ducks). Considering the spline as a beam and the ducks as concentrated loads on the beam, then the deformed beam satisfies the Bernoulli-Euler equation

$$EIk(x) = M(x), \tag{3.4.1}$$

where EI is a constant measures rigidity, $M(x)$ is the bending moment and $k(x)$ denotes the curvature of the beam. We further suppose that $M(x)$ is a linear function of x. Letting $\sigma(x)$ denote the arc represented by the bent beam, then the curvature is given by

$$k(x) = \frac{\sigma''}{(1+\sigma'^2)^{3/2}}.$$

We need the following assumptions.

Assumption 1: the beam deflections are small, the tangent vector satisfies $\|\sigma'\| \ll 1$. Equivalently stated, σ' is negligible.

Hence, we may suppose that $k(x) = (\sigma'')$. Since $M'' = 0$, then the fourth derivative $\sigma^{(4)} = 0$. These leads to the following.

Assumption 2: σ is twice continuously differentiable at each duck and cubic between the ducks.

For notational simplicity, we write this development as if $\sigma : [a, b] \to \mathbb{R}$. Nevertheless, there is no material difference for vector valued functions.

We are now in a position to solve for σ. In particular, we suppose that the physical spline has pins with x-coordinates at each x_i, $a = x_0 < x_1 < \dots < x_n = b$. Furthermore, we require $\sigma = \sigma_i$ on $[x_{i-1}, x_i]$, where each σ_i is a parametric cubic, and σ is twice continuously differentiable at each x_i. We set $\sigma'(x_i) = m_i$ and $\sigma''(x_i) = M_i$.

Since the second derivative of σ_i is linear, then we have

$$\sigma_i''(x) = \frac{M_{i-1}(x_i - x)}{\Delta_i} + \frac{M_i(x - x_{i-1})}{\Delta_i}, \tag{3.4.2}$$

where $\Delta_i = x_i - x_{i-1}$. Indeed, the right hand side of (3.4.2) is the unique linear function which is M_{i-1} at x_{i-1} and M_i at x_i. Integrating (3.4.2) twice we get successively

$$\sigma_i'(x) = \frac{M_{i-1}(x_i - x)^2}{2\Delta_i} + \frac{M_i(x - x_{i-1})^2}{2\Delta_i} + C,$$

$$\sigma_i(x) = \frac{M_{i-1}(x_i - x)^3}{6\Delta_i} + \frac{M_i(x - x_{i-1})^3}{6\Delta_i} + Cx + D. \tag{3.4.3}$$

Evaluating (3.4.3) at x_{i-1} and x_i we get

$$y_{i-1} = \frac{M_{i-1}(x_i - x_{i-1})^3}{6\Delta_i} + Cx_{i-1} + D,$$

$$y_i = \frac{M_i(x_i - x_{i-1})^3}{6\Delta_i} + Cx_i + D,$$

where we write $y_i = \sigma(x_i)$. Subtracting y_{i-1} from y_i, we have

$$y_i - y_{i-1} = \frac{M_i(x_i - x_{i-1})^3}{6\Delta_i} - \frac{M_{i-1}(x_i - x_{i-1})^3}{6\Delta_i} + C(x_i - x_{i-1}).$$

Dividing through by $\Delta_i = x_i - x_{i-1}$ yields,

$$\frac{y_i - y_{i-1}}{\Delta_i} = (M_i - M_{i-1})\frac{\Delta_i}{6} + C.$$

Now solving for C and substituting into (3.4.3) we get

$$\sigma_i'(x) = M_{i-1}\frac{(x_i - x)^2}{2\Delta_i} + M_i\frac{(x - x_{i-1})^2}{2\Delta_i}$$

$$+\frac{y_i - y_{i-1}}{\Delta_i} - (M_i - M_{i-1})\frac{\Delta_i}{6}$$

Using similar but significantly more complicated calculations we get

$$D = \frac{y_{i-1}}{\Delta_i}x_i + M_{i-1}\frac{\Delta_i}{6}x_i - \frac{y_i}{\Delta_i}x_{i-1} + M_i\frac{\Delta_i}{6}x_{i-1}.$$

Hence, we have

$$\sigma_i(x) = \frac{M_{i-1}(x_i - x)^3}{6\Delta_i} + \frac{M_i(x - x_{i-1})^3}{6\Delta_i} + \frac{y_i - y_{i-1}}{\Delta_i}$$

$$-(M_i - M_{i-1})\frac{\Delta_i}{6}x + \frac{y_{i-1}}{\Delta_i}x_i + M_{i-1}\frac{\Delta_i}{6}x_i - \frac{y_i}{\Delta_i}x_{i-1} + M_i\frac{\Delta_i}{6}x_{i-1}. \tag{3.4.4}$$

Note that the unknowns in equation (3.4.4) are M_{i-1} and M_i, and that the equations are linear in these two variables. That is, (3.4.4) describes a linear system of equations. We now organize this equation to get an expression for the second derivatives (the M_i) without reference to x, the independent variable, and then write the resulting linear system in matrix form.

First, we write (3.4.4) in a slightly more compact form,

$$\sigma_i(x) = \frac{M_{i-1}(x_i - x)^3}{6\Delta_i} + \frac{M_i(x - x_{i-1})^3}{6\Delta_i}$$

$$+(\frac{y_i}{\Delta_i} - M_i\frac{\Delta_i}{6})(x - x_{i-1}) + (\frac{y_{i-1}}{\Delta_i} - M_{i-1}\frac{\Delta_i}{6})(x_i - x).$$

And for σ_{i+1} we get

$$\sigma_{i+1}(x) = \frac{M_i(x_{i+1} - x)^3}{6\Delta_{i+1}} + \frac{M_{i+1}(x - x_i)^3}{6\Delta_{i+1}}$$

$$+(\frac{y_{i+1}}{\Delta_{i+1}} - M_{i+1}\frac{\Delta_{i+1}}{6})(x - x_i) + (\frac{y_i}{\Delta_{i+1}} - M_i\frac{\Delta_{i+1}}{6})(x_{i+1} - x).$$

Now we differentiate these last two expressions with respect to x, evaluate the derivative at x_i and use $\sigma_i'(x_i) = \sigma_{i+1}'(x_i)$ to get

$$M_i\frac{\Delta_i}{2} + \frac{y_i}{\Delta_i} - M_i\frac{\Delta_i}{6} - \frac{y_{i-1}}{\Delta_i} + M_{i-1}\frac{\Delta_i}{6}$$

$$= -M_i\frac{\Delta_{i+1}}{2} + \frac{y_{i+1}}{\Delta_{i+1}} - M_{i+1}\frac{\Delta_{i+1}}{6} - \frac{y_i}{\Delta_{i+1}} + M_i\frac{\Delta_{i+1}}{6}.$$

Now separating the terms, we have

$$M_{i-1}\frac{\Delta_i}{6} + M_i\frac{\Delta_i}{3} + M_i\frac{\Delta_{i+1}}{3} + M_{i+1}\frac{\Delta_{i+1}}{6} = \frac{y_{i+1} - y_i}{\Delta_{i+1}} - \frac{y_{i+1} - y_i}{\Delta_i}.$$

Next, we multiply by 6 and divide by $(\Delta_{i+1} + \Delta_i)$,

$$\frac{\Delta_i}{\Delta_{i+1} + \Delta_i}M_{i-1} + 2M_i + \frac{\Delta_i}{\Delta_{i+1} + \Delta_i}M_{i+1}$$

$$= \frac{6}{\Delta_{i+1} + \Delta_i}\Big(\frac{y_{i+1} - y_i}{\Delta_{i+1}} - \frac{y_{i+1} - y_i}{\Delta_i}\Big). \qquad (3.4.5)$$

Equation (3.4.5) represents a linear system of equations. There are $n + 1$ unknowns, M_j, $j = 0, ..., n$ and $n - 1$ equations associated to the spline segments, σ_i, $i = 1, ..., n-1$. Setting boundary values at M_0 and M_n adds two more equations to the system,

$$M_0 = \delta_0, \;\; M_n = \delta_n. \tag{3.4.6}$$

If we denote the right hand side of (3.4.5) by d_i and set $\Delta_i/(\Delta_{i+1}+\Delta_i) = \lambda_i$, $\Delta_{i+1}/(\Delta_{i+1}+\Delta_i) = \mu_i$, then (3.4.5) and (3.4.6) yield the following lower triangular linear system in matrix form.

$$\begin{pmatrix} 1 & 0 & 0 & \dots & 0 & 0 & 0 & 0 \\ 2 & \lambda_2 & 0 & \dots & 0 & 0 & 0 & 0 \\ \mu_2 & 2 & \lambda_3 & \dots & 0 & 0 & 0 & 0 \\ 0 & \mu_3 & 2 & \dots & 0 & 0 & 0 & 0 \\ \dots & \dots & \dots & \dots & \dots & \dots & \dots & \dots \\ 0 & 0 & 0 & \dots & \mu_{n-3} & 2 & \lambda_{n-2} & 0 \\ 0 & 0 & 0 & \dots & 0 & \mu_{n-2} & 2 & \lambda_{n-1} \\ 0 & 0 & 0 & \dots & 0 & 0 & \mu_{n-1} & 2 \\ 0 & 0 & 0 & \dots & 0 & 0 & 0 & 1 \end{pmatrix} \begin{pmatrix} M_0 \\ M_1 \\ M_2 \\ M_3 \\ \dots \\ M_{n-3} \\ M_{n-2} \\ M_{n-1} \\ M_n \end{pmatrix} = \begin{pmatrix} \delta_0 \\ d_1 \\ d_2 \\ d_3 \\ \dots \\ d_{n-3} \\ d_{n-2} \\ d_{n-1} \\ \delta_n \end{pmatrix} \tag{3.4.7}$$

The solution to (3.4.7) is a *piecewise polynomial function.* It is not a polynomial, but rather the join of polynomial segments. Further, the spline curve interpolates the given set of points. Hence, we have a process similar to polynomial interpolation. However, based on the physical model, splines will have local control. In particular, the problem that we identified with polynomial interpolation cannot happen here. On the negative side, the size of the linear system in (3.4.7) depends on the number of knot points or *spline guide points.* We encountered the same situation with polynomial interpolation.

There is an alternative called the B-spline. B-splines are piecewise cubics that exhibit local control and do not require that we solve a linear system. On the downside, they do not interpolate the guide points. Credit for these curves is given to I. J. Schoenberg [Schoenberg (1973)].

The basic idea is to remove the physical context from the spline while keeping the basic properties. In particular, we want a piecewise parametric cubic polynomial that is twice continuously differentiable at the knot points. In particular, suppose we have a list of distinct points, $P_i = (x_i, y_i), i = 0, ..., n$. Suppose further that $x_i < x_{i+1}$. We seek a function σ defined on an interval $[a, b]$, $x_0 \le a < b \le x_n$ with the following properties. The function σ is the join of several segments, σ_k. Each segment, σ_k, is a parametric cubic defined on $[0, 1]$ with $\sigma_k^{(j)}(1) = \sigma_{k+1}^{(j)}(0)$ for $j = 0, 1, 2$. In short, the function σ is locally a cubic polynomial and globally twice continuously differentiable.

We carry this line of thought a step further. We know that $\mathcal{P}_3$, the set of all polynomials of degree less than or equal to three, is a four dimensional vector space. We want to know if $\mathcal{P}_3$ have a basis p_1, p_2, p_3, p_4 with the following property. For any x_1, x_2, x_3, x_4 and x_5 in $\mathbb{R}$,

$$x_1 p_1^{(j)}(1) + x_2 p_2^{(j)}(1) + x_3 p_3^{(j)}(1) + x_4 p_4^{(j)}(1)$$

$$= x_2 p_1^{(j)}(0) + x_3 p_2^{(j)}(0) + x_4 p_3^{(j)}(0) + x_5 p_4^{(j)}(0), \qquad (3.4.8)$$

for derivatives $j = 0, 1, 2$. If the answer is affirmative, then we may set $\sigma_1(t) = \sum_i x_i p_i(t)$ and $\sigma_2(t) = \sum_i x_{i+1} p_i(t)$ and join σ_1 and σ_2 to form a piecewise cubic, twice continuously differentiable function spanning between

$$\frac{1}{6}x_1 + \frac{4}{6}x_2 + \frac{1}{6}x_3, \quad \frac{1}{6}x_3 + \frac{4}{6}x_4 + \frac{1}{6}x_5.$$

If we set $p_i(X) = \sum_{k=0}^{3} \alpha_{ik} X^k$, then the three constraints given in (3.4.8) become

$$\sum_{i=1}^{4} (x_i - x_{i+1})\alpha_{i0} + x_i \alpha_{i1} + x_i \alpha_{i2} + x_i \alpha_{i3} = 0 \qquad (3.4.9)$$

$$\sum_{i=1}^{4} (x_i - x_{i+1})\alpha_{i1} + 2x_i \alpha_{i2} + 3x_i \alpha_{i3} = 0 \qquad (3.4.10)$$

$$\sum_{i=1}^{4} (x_i - x_{i+1})\alpha_{i2} + 3x_i \alpha_{i3} = 0 \qquad (3.4.11)$$

Since these equations may hold for any x_i, we set $x_1 = x_2 = x_3 = x_4 = 0$ and $x_5 = 1$ in (3.4.9) to yield $\alpha_{40} = 0$. In turn (3.4.10) implies that $\alpha_{41} = 0$ and (3.4.11) implies that $\alpha_{42} = 0$. Therefore, $p_4(X) = \alpha_{43} X^3$. Continuing in this manner, we derive $p_1(X) = \alpha_{40}(-X^3 + 3X^2 - 3X + 1)$, $p_2(X) = \alpha_{41}(3X^3 - 6X^2 + 4)$, $p_3(X) = \alpha_{42}(-3X^3 + 3X^2 + 3X + 1)$. (See Exercise 6.) We normalize by setting each $\alpha_{4i} = 1/6$. With this choice, $\sum_i p_i = 1$. These four polynomials are called the *B-spline basis functions*.

These B-spline basis functions may be used to define a powerful fitting process. Given a list of points, $P_i = (x_i, y_i)$, $i = 1, ..., n$, we write,

$$\sigma_j(t) = \sum_{i=1}^{4} p_i(t) P_{j+i-1} = \frac{1}{6}\big[(-t^3 + 3t^2 - 3t + 1)P_j + (3t^3 - 6t^2 + 4)P_{j+1}$$

$$+(-3t^3 + 3t^2 + 3t + 1)P_{j+2} + t^3 P_{j+3}\big], \qquad (3.4.12)$$

where $t \in [0, 1]$ and $j \leq n - 3$. We have the following theorem.

Theorem 3.4.1. *Given points $P_i = (x_i, y_i)$, $i = 1, ..., n$, the function $\sigma(t)$ defined by joining the segments given by (3.4.12) is a parametric function that is piecewise cubic and twice continuously differentiable.*

Proof. The result follows from the properties of the B-spline basis functions. □

We call σ the *B-spline fit* for the given point set. Keep in mind that σ does not extend from P_1 to P_n but spans between the points

$$\frac{1}{6}P_1 + \frac{4}{6}P_2 + \frac{1}{6}P_3, \ \frac{1}{6}P_{n-2} + \frac{4}{6}P_{n-1} + \frac{1}{6}P_n.$$

We carry this a step further. Consider a function f with values $f(x_i) = y_i$, and set $\Delta = \max_i \Delta_i$. Suppose further that f is *Lipschitz continuous* in the sense that $|f(x) - f(y)| < K|x - y|$ for any x, y in the domain of f and a constant K depending only on f. With this notation and terminology, we state the following theorem. The proof requires knowledge of real analysis. It is left as an exercise.

Theorem 3.4.2. *Suppose f is a Lipschitz continuous function defined on a real interval $[a, b]$. Further, suppose that $a = x_0 \leq x_1 \leq ... \leq x_n = b$ is a partition of the domain denoted by $\mathcal{P}$. We take $P_i = (x_i, f(x_i))$ and write $\sigma_{\mathcal{P}}$ for the associated B-spline fit. It follows that $\sigma_{\mathcal{P}} \to f$ order Δ.*

Proof. See Exercise 7. □

In particular, the B-spline fit for the function is not an interpolation but it does approximate the function.

Before proceeding note that a continuously differentiable function on a closed interval is Lipschitz continuous. See Exercise 8.

Finally, we consider *local control* for B-splines. First, note that each point on a B-spline segment σ_i is a linear combination of the four guide points for the segment. Furthermore, as the sum of the B-spline basis functions is 1, then sum of the coefficients in the linear combination is also 1. Hence, the curve lies within the convex hull of the four guide points. Note that the convex hull of a point set is the smallest convex set that contains the points. For 4 points, it is the convex quadrilateral or triangle

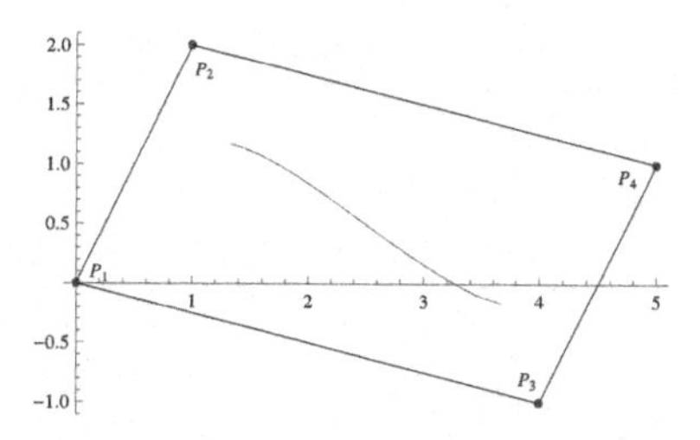

Figure 3.4.1: The B-spline segment and the convex hull

that contains the points. The following diagram illustrates the convex hull property. The spline segment is generated by the four points.

In summary, B-splines are parametric cubic curves which fit the given set of points, guide points. Each B-spline segment is determined by four of the guide points. For instance, segment σ_j is determined by the four guide points, $P_j, P_{j_1}, P_{j+2}, P_{j+3}$. In practice each segment traces a curve very close to its four determining guide points. The B-spline curve (the join of the segments) is twice continuously differentiable. This high level of smoothness is often the reason that B-splines are used.

In addition, with the convex hull property you can predict the effect on the curve that arises from the change in one of the guide points. Any given guide point is included in the calculation of at most four curve segments, or equivalently at most four of the bounding convex hulls. Therefore, if you change a guide point, then you can predict the change in the curve by looking at how the convex hulls change. Figures 3.4.2 shows a two segment B-spline. The convex hull for the first segment is formed by P_1, P_2, P_3, P_4. For the second segment the convex hull is determined by P_2, P_3, P_5, P_4. In Figure 3.4.3, we have moved P_3.

Smoothness, local control and convergence are strong advantages for B-splines over polynomials. Since there is no need to solve a large linear system, B-splines are preferred over general cubic splines. B-splines are smooth, piecewise Bezier curves are not. Indeed, unless exact interpolation is absolutely required, B-splines are often the technique of choice. Finally, computer aided design software and medical imaging software often represent objects using B-spline curves and surfaces. Hence, subsequent analysis will be B-spline based.

Exercises:

1. Plot the four B-spline basis functions.

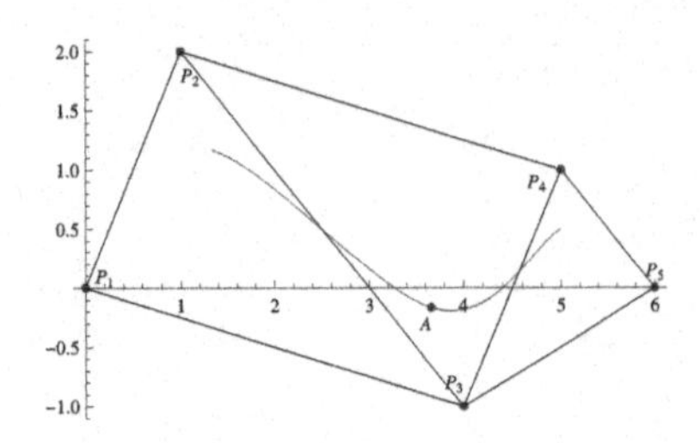

Figure 3.4.2: Two B-spline segments with knot point A

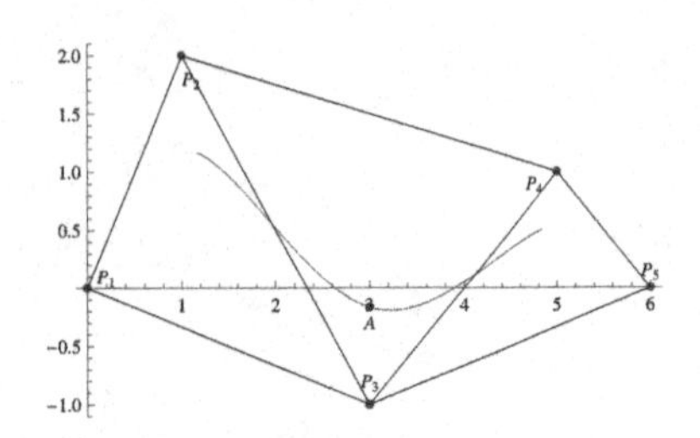

Figure 3.4.3: Two B-splines, P_3 has been moved to the left

2. Compute the B-spline fit to the points $(x_i, 1/(1+x_i^2)$ for $x_i = -4, -2, -1, 0, 1, 2, 4$. Plot the parametric B-spline against the graph of $f(x) = 1/(1+x^2)$. Compare your output with Exercise 3 of Section 3.1.

3. In Exercise 3, the fourth guide point is (0, 1). Leaving the other guide points unchanged, repeat Exercise 3 using (0, 1.1) for the fourth guide point. Compare the output to the curve in Exercise 2.

4. Repeat Exercise 3 with fourth guide point equal to (0, 1.2) or (0.0.9) or (0, 0.8).

5. Repeat Exercise 7 of Section 3 with the same data. This time fit B-Splines to the data and plot the curve on the same graph with the data. Use the B-Spline to predict values for $f(750)$. $f(850)$ and $f(950)$. Determine the mean absolute error and compare the result to the result using least squares fitting. Which method was better?

6. Complete the derivation of the B-Spline basis functions.

7. Prove Theorem 3.4.2 by following the given outline.
a. Determine the $\max_{t\in[0,1]} |p_i(t)|$ for the four B-spline basis functions.
b. Verify that $\sum_{i=1}^{4} p_i(x) = 1$.
c. Let $\sigma_i(t) = (\sigma_{i,1}(t), \sigma_{i,2}(t))$ be a B-spline segment for $\sigma_{\mathcal{P}}$ and take $t_0 \in [0,1]$. Prove that if $x_0 = \sigma_{i,1}(t_0)$, then

$$|f(x_0) - \sigma_{i,2}(t_0)| \leq \sum_{k=0}^{3} |p_{i+k}(t_0)||f(x_0) - f(x_{i+k})|.$$

d. Complete the proof of Theorem 3.4.2 by identifying a constant C depending on f that satisfies

$$|f(x_0) - \sigma_{i,2}(t_0)| \leq C\Delta.$$

8. Prove that for any continuously differentiable function f on a closed interval, f' is bounded. Conclude that f is Lipschitz continuous.

3.5 Hermite Interpolation

Hermite interpolation goes one step beyond polynomial interpolation. In Section 3.1, we considered a continuous function f and a finite set of interpolation points x_i, $i = 0, 1, ..., n$. In this setting we determined a polynomial p that matched f at the given points. Now we assume that f is differentiable and we require that $f(x_i) = p(x_i)$ and $f'(x_i) = p'(x_i)$ at each of the interpolation points. The resulting polynomial will be called the *Hermite interpolation*

We will see in the Chapter 5 that Hermite interpolation gives rise to Gaussian quadrature, the generally accepted *gold standard* of numerical integration techniques. But that is just one of the applications. In the case of piecewise interpolation (introduced in Section 3.1) the resulting interpolant will be C^2. This has implications for finite element method in one spatial dimension for rectangular higher dimensional domains. In another direction, the need to model both the function and its derivative leads to discontinuous Galerkin method. There are at least two more application of Hermite interpolation in the area of numerical differential equations. One is OSC or orthogonal spline collocation and the other is *Hermitian method*, an interesting technique designed specifically for non-linear equations in one spatial dimension.

We begin with the formal definition.

Definition 3.5.1. Suppose that f is a differentiable function on $[a, b]$ and take $a \leq x_0 < x_1 < ... < x_{n-1} < x_n \leq b$. The expression

$$h(x) = \sum_{i=0}^{n} f(x_i)H_i(x) + \sum_{i=0}^{n} f'(x_i)S_i(x)$$

is a *Hermite interpolation* of f provided the polynomials H_i and S_i are degree $2n + 1$ and satisfy,

(i) $H_i(x_j) = \delta_{i,j}$, $S_i(x_j) = 0$

(ii) $dH_i/dx(x_j) = 0$, $dS_i/dx(x_j) = \delta_{i,j}$.

Our first task is to show that the Hermite interpolation exists; that is, polynomials exist that satisfy the conditions of definition 4.4.1. We start with the $n+1$, n^{th} degree Lagrange polynomials associated to the points x_i, $i = 0, .., n$. Recall from Section 3.1,

$$l_i(x) = \frac{\prod_{j\neq i}(x - x_i)}{\prod_{j\neq i}(x_j - x_i)}.$$

Since $l_i(x_j) = \delta_{i,j}$, then the same is true for $h_i = l_i^2$. Furthermore, if we set $p(x) = \prod_{i=0}^{n}(x - x_i)$, then

$$p'(x) = \sum_i \prod_{j\neq i}(x - x_j); \;\; p'(x_i) = \prod_{j\neq i}(x_i - x_j).$$

Hence, $l_i(x) = p(x)/[(x - x_i)p'(x_i)]$. We compute

$$\frac{d}{dx}h_i = 2l_i(x)\frac{d}{dx}l_i(x)$$

$$= 2\frac{1}{x - x_i}\frac{p(x)}{p'(x_i)}\left(\frac{-1}{(x - x_i)^2}\frac{p(x)}{p'(x_i)} + \frac{1}{x - x_i}\frac{p'(x)}{p'(x_i)}\right).$$

Therefore, for $j \neq i$, $h_i^{'}(x_j) = 0$.

At this point we have $n+1$ functions h_i, which are polynomials of degree $2n$. Next, we seek polynomials u_i and v_i satisfying the following,

a. degree $u_i = 1$, $u_i(x_i) = 1$, $u_i^{'}(x_i) = -h_i^{'}(x_i)$;

b. degree $v_i = 1$, $v_i(x_i) = 0$, $v_i^{'}(x_i) = 1$.

Indeed, in this case $H_i = u_i h_i$ and $S_i = v_i h_i$ satisfy the conditions of Definition 4.1.1. It is immediate that $v_i(x) = x - x_i$ is consistent with item b above. In turn, we set $u_i(x) = \alpha(x - x_i) + 1$ where $\alpha = -h_i^{'}(x_i)$. It is now a simple matter to complete the proof of the following theorem.

Theorem 3.5.1. *Any differentiable function on an interval has Hermite interpolation.*

With existence in hand, the next task is to estimate the error.

Theorem 3.5.2. *Suppose that f is $2n+2$ times continuously differentiable on (a, b) and that h is the Hermite interpolation of f with respect to a partition of $[a, b]$. If we denote the error, $e(x) = f(x) - h(x)$, then for each $\hat{x}$, there is $\xi_{\hat{x}}$ in the interval with*

$$e(\hat{x}) = \frac{f^{(2n+2)}(\xi_{\hat{x}})}{(2n+2)!}p(\hat{x})^2. \tag{3.5.1}$$

If $M = \|f^{(2n+2)}\|_\infty$, *then*

$$|e(\hat{x})| \leq \frac{M}{(2n+2)!} p(\hat{x})^2. \tag{3.5.2}$$

Furthermore, if f is a polynomial of degree less than or equal to $2n+1$, *then* $f = h$.

Proof. The proof of this is analogous to the derivation of the error estimate for polynomial interpolation as developed in Section 3.1. As in that case, it is an application of Rolle's theorem. We will set up the argument and leave the final details to the reader.

To begin, we set

$$e(x) = f(x) - h(x); \quad g(x) = \frac{e(x)}{p(x)^2}.$$

Next, we take ζ distinct for any of the x_i and define

$$k(x) = f(x) - h(x) - p(x)^2 g(\zeta).$$

Now, k has $n+2$ distinct roots at the x_i and at ζ. By Rolle's theorem, k' has $n+1$ distinct roots each distinct from the roots of k. In addition, since h is the Hermite interpolation, the x_i are also roots of k'. This gives us at least $2n+2$ distinct roots for k'. Therefore, $k^{(2n+2)}$ has at least one root. Now complete the proof as in Theorem 3.1.3.

Equation (3.5.2) is an easy consequence of (3.5.1). The final assertion is immediate. □

There is a special case of Hermite interpolation that is too important to omit. The following 4 polynomials form a basis for the degree 3 polynomials on $[\alpha, \beta]$

$$H_0 = 3\left(\frac{\beta - x}{\beta - \alpha}\right)^2 - 2\left(\frac{\beta - x}{\beta - \alpha}\right)^3, H_1 = 3\left(\frac{x - \alpha}{\beta - \alpha}\right)^2 - 2\left(\frac{x - \alpha}{\beta - \alpha}\right)^3,$$

$$S_0 = \frac{(\beta - x)^2}{\beta - \alpha} - \frac{(\beta - x)^3}{(\beta - \alpha)^2}, S_1 = -\frac{(x - \alpha)^2}{\beta - \alpha} + \frac{(x - \alpha)^3}{(\beta - \alpha)^2}.$$

These polynomials are called the *Hermite cubic polynomials.* The function H_0 satisfies, $H_0(\alpha) = 1, H_0(\beta) = 0$ and both derivatives of H_0 at the end points are zero. In addition, S_0 vanishes at the end points, and the slope at α is 1 and the slope at β is zero. H_1 and S_1 have analogous properties.

It is not difficult to prove that the 4 Hermite cubics are linear independent. Indeed, if $aH_0 + bH_1 + cS_0 + dH_1 = 0$ then

$$0 = aH_0(\alpha) + bH_1(\alpha) + cS_0(\alpha) + dH_1(\alpha) = aH_0(\alpha) = a.$$

Similarly, $b = c = d = 0$. We use a dimension argument to conclude that the four Hermite cubics form a basis for the degree three polynomials.

These degree three polynomials are commonly used to model the solution to a second order PDE.

Exercises

1. Verify that when n is 1, $a = x_0$ and $b = x_1$, then the Hermite cubics provide a Hermite interpolation of $[a, b]$.

2. Complete the verification that the functions $u_i h_i$ and $v_i h_i$ satisfy the requirements for a Hermite interpolation.

3. Complete the proof of Theorem 3.5.1.

4. Verify that the Hermite cubics determine a Hermite interpolation on the interval $[\alpha, \beta]$.

5. Plot the 4 Hermite cubics and verify by inspection that they satisfy the requirements of Definition 3.5.1.

Chapter 4

Numerical Differentiation

Introduction

In this chapter, we begin the study of the basic constructs of calculus from the standpoint of numerical analysis. In particular, we look at procedures to approximate the derivative of a function when we only know values of the function but have no closed form or series representation. In the next chapter we consider the integral.

We begin by introducing the finite difference as the basic technique to approximate the derivative of a function from only knowledge of some of the function values. For instance, if we know values of our function f at points a and b, then $(f(b) - f(a))/(b - a)$ is approximately equal to the derivative provided $b - a$ is small. From the numerical analysis point of view, we want an error estimate for any approximation. If we suppose that f is sufficiently smooth, then we can use the Taylor expansion to derive the finite difference formulas. In this case, Taylor's theorem provides the error estimate.

Finite differences provide the entry point into numerical processes to approximate the solution of a differential equation. In the second section, we introduce the finite difference method (FDM) to estimate solutions of the one-dimensional heat equation. In this section we develop the explicit FDM formulation. On one hand the technique is successful. However, almost immediately, we see that this technique can give rise to unacceptable levels of error. This leads us to consider von Neumann stability. This is the subject of Section 3. In Section 4, we introduce implicit and Crank Nicolson FDM. We prove immediately that stability is not an issue for these procedures. Crank Nicolson is often the method of choice for parabolic partial differential equations (PDE). In the exercises we consider FDM as applied to the order 1 wave equation. We learn that FDM applied to hyperbolic PDE is very different. Little of the intuition gained from the parabolic case can be brought forward to the hyperbolic equations.

4.1 Finite Differences and Vector Fields

If we start with a function, f which is real valued and differentiable at x_0, then the derivative is given by $f'(x_0) = \lim_{x \to x_0}(f(x) - f(x_0))/(x - x_0)$. Hence, for x near to x_0, the derivative at x_0 is approximately equal to $(f(x) - f(x_0))/(x - x_0)$. This is the basic idea behind the finite difference.

Finite differences are useful because the researcher often knows values of a function without specifically knowing the function. For instance, the researcher observes a process, measures values $(x, f(x))$, but has formula for f.

With finite differences, the researcher can infer approximate values of the derivative of an unknown function. We will see an example of this later in the section. Furthermore, we see in subsequent sections that finite differences provide a valuable tool for approximating the solution of a differential equation. We turn now to the formal definition.

Let f be real valued function defined on an interval $[a, b]$. For a positive integer n set $\Delta x = (b - a)/n$ and consider a uniform partition $a = x_0 < x_1 < ... < x_n = b$, where for each i, $x_i - x_{i-1} = \Delta x$. Suppose further that f is four times continuously differentiable, so that we may write the order three Taylor expansion for f at x_i evaluated at x_{i+1}.

$$f(x_{i+1}) = f(x_i) + \frac{df}{dx}(x_i)\Delta x + \frac{1}{2}\frac{d^2 f}{dx^2}(x_i)\Delta x^2 + \frac{1}{6}\frac{d^3 f}{dx^3}(x_i)\Delta x^3 + \mathcal{O}(\Delta x^4). \tag{4.1.1}$$

The term $\mathcal{O}(\Delta x^4)$ is the Taylor remainder term. In particular, this term converges to zero order 4 as Δx goes to zero.

In turn, we have

$$f(x_{i-1}) = f(x_i) - \frac{df}{dx}(x_i)\Delta x + \frac{1}{2}\frac{d^2 f}{dx^2}(x_i)\Delta x^2 - \frac{1}{6}\frac{d^3 f}{dx^3}(x_i)\Delta x^3 + \mathcal{O}(\Delta x^4). \tag{4.1.2}$$

In (4.1.1), subtracting $f(x_i)$ from both sides and dividing by Δx yields

$$\begin{aligned} \frac{f(x_{i+1}) - f(x_i)}{\Delta x} &= \frac{df}{dx}(x_i) + \frac{1}{2}\frac{d^2 f}{dx^2}(x_i)\Delta x + \frac{1}{6}\frac{d^3 f}{dx^3}(x_i)\Delta x^2 + \mathcal{O}(\Delta x^3) \\ &= \frac{df}{dx}(x_i) + \mathcal{O}(\Delta x). \end{aligned} \tag{4.1.3}$$

If we do the same to (4.1.2) we obtain

$$\frac{f(x_i) - f(x_{i-1})}{\Delta x} = \frac{df}{dx}(x_i) + \mathcal{O}(\Delta x). \tag{4.1.4}$$

Note that we only need f to be C^2 to derive Equations (4.1.3) and (4.1.4).

We state formally,

Definition 4.1.1. Consider a differentiable real valued function f defined on the interval $[a, b]$ with uniform partition $a = x_0 < x_1 < \ldots < x_n = b$ and $\Delta x = x_i - x_{i-1}$. The *first forward difference* and *first backward difference* for f are given by

$$\frac{f(x_{i+1}) - f(x_i)}{\Delta x}; \quad \frac{f(x_i) - f(x_{i-1})}{\Delta x}.$$

The following theorem is an immediate consequence of (4.1.3) and (4.1.4). We continue the current notation.

Theorem 4.1.1. *If f is twice continuously differentiable on $[a, b]$ then the first forward and backward differences at x_i converge to $f'(x_i)$ order Δx.*

Next, we add (4.1.1) to (4.1.2) and divide by $2\Delta x$ to get

$$\begin{aligned}\frac{f(x_{i+1}) - f(x_{i-1})}{2\Delta x} &= \frac{df}{dx}(x_i) + \frac{1}{6}\frac{d^3 f}{dx^3}(x_i)\Delta x^2 + \mathcal{O}(\Delta x^3)\\ &= \frac{df}{dx}(x_i) + \mathcal{O}(\Delta x^2).\end{aligned}$$

In turn, if we subtract (4.1.2) from (4.1.1) and divide by Δx^2 we get

$$\frac{f(x_{i+1}) - 2f(x_i) + f(x_{i-1})}{\Delta x^2} = \frac{d^2 f}{dx^2}(x_i) + \mathcal{O}(\Delta x^2). \tag{4.1.5}$$

These two equations define the central differences approximating the first and second derivatives of f.

Definition 4.1.2. Consider a three times differentiable real valued function f defined on the interval $[a, b]$ with uniform partition $a = x_0 < x_1 < \ldots < x_n = b$ and $\Delta x = x_i - x_{i-1}$. The *first central difference* and *second central difference* for f are given by

$$\frac{f(x_{i+1}) - f(x_{i-1})}{2\Delta x}; \quad \frac{f(x_{i+1}) - 2f(x_i) + f(x_{i-1})}{\Delta x^2}.$$

Equations (4.1.5) and (4.1.6) resolve the order of convergence for the central differences. It is important to note that the central differences are better behaved than the forward and backward differences. We revisit this in the exercises.

Theorem 4.1.2. *If f is three times continuously differentiable on $[a, b]$ then the first central and second central differences at x_i converge to $f'(x_i)$ and $f''(x_i)$ order Δx^2.*

When defining the finite differences, we used a uniform partition along the x-axis. Generally speaking, finite differences perform best when the partition is uniform. But there are many other forms of the finite difference which are used in particular circumstances. With the aid of these more advanced techniques applied to certain special cases, researchers can achieve good results with non-uniform partitions. For our purposes here, we will restrict attention to the uniform case. For a more complete treatment, see [Loustau (2016)].

Next, consider the following setting from computational fluid dynamics. Suppose we have a partially obstructed channel as shown in Figure 4.1.1. The upper and lower boundaries are the channel edges, the right and left margins are the inflow and outflow edges. We will suppose that the fluid moves from left to right. Using results from complex analysis, there is a real valued flow potential function φ defined on the region in D in Figure 4.1.1. In particular, the potential satisfies the Laplace equation, $\nabla^2 \varphi = 0$. In this case, $\nabla \varphi = (\varphi_x, \varphi_y)$ defines the vector field that represents a fluid flow through the channel. By the Riemann mapping theorem, we know that φ exists, but have no means to solve for it. Nevertheless, using advanced numerical techniques, we can approximate this function at locations in D. We do this for a uniform grid (x_i, y_j), $x_{i+1} - x_i = \Delta x$, $y_{j+1} - y_j = \Delta y$. We show the grid in Figure 4.1.2. There are about 230 locations in the grid. Next, we use forward differences to approximate the vector field,

$$\left(\frac{\varphi(x_{i+1}, y_j) - \varphi(x_i, y_j)}{\Delta x}, \frac{\varphi(x_i, y_{j+1}) - \varphi(x_i, y_j)}{\Delta y}\right)$$

The resulting vector field (see Figure 4.1.3) illustrates the fluid flow. We used the *Arrow* function in *Mathematica* graphics to display the output. The length of the vectors represents the speed of the flow at the location.

In Exercise 2, we use finite differences to approximate the flow field about a cylinder. In this case, we know the flow potential function. Hence, for this example we can compare the finite difference approximations to the actual.

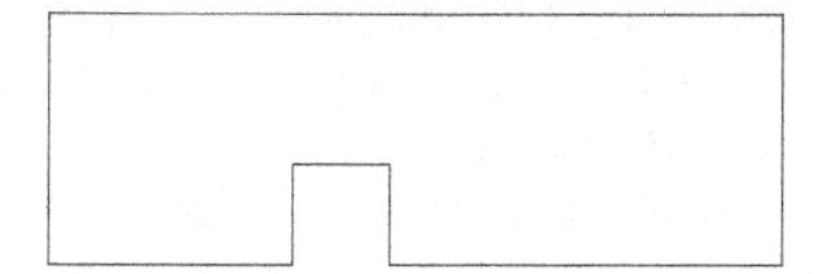

Figure 4.1.1: The partially obstructed channel

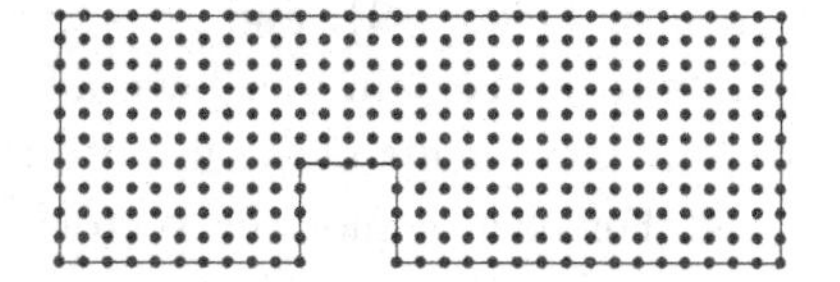

Figure 4.1.2: The channel with uniform grid in x and y directions

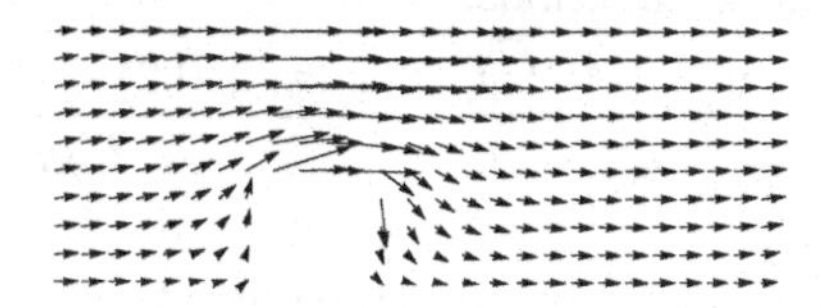

Figure 4.1.3: The flow field approximated with finite differences

There is an alternative direction described in Exercise 3. In this case, we know the complex potential function but cannot get the real part and compute the gradient. Finite differences provide a means to do this computation.

Exercises:

1. Take $f(x) = x^2$ and $\Delta x = 0.1$.

a. Compute the first forward, backward and central differences at $x = 1, 2, ..., 100$. Compare the ouput against the actual derivative. Compute the mean absolute error for each case.

b. Which finite difference procedure is better? Why?

c. Compute the second central difference at the given locations. Compare the output against the actual second derivatives and compute the mean absolute error.

2. Consider the following function, $u(x, y) = x + x/(x^2 + y^2)$ defined on the domain $D = \{(x, y) : -2 \leq x \leq 2,\ 0 \leq y \leq 2,\ x^2 + y^2 \geq 1\}$. We can think of this domain as a partially obstructed channel. In this case the obstruction is a semi-circle.

a. Graph the domain D.

b. Create a list of points in D so that $P_i = (x_i, y_i)$ is in the list provided $x_i = -2 + 0.2m$ and $y_i = 0.2n$ for some positive integers m in $\{1, 2, ..., 19\}$

and n in $\{1, 2, ..., 9\}$, and $x^2 + y^2 > 1$. Before proceeding plot the list of points to be certain you have all points in D whose x and y coordinates step by increments of 0.2 in both the x and y direction.

c. For each i, plot the vector $(x_i, y_i) + \nabla u(x_i, y_i)$. Use the arrowhead feature when drawing the vectors. You should see the vector field describing the flow of a non-viscous, incompressible fluid around a semi-circular obstruction.

d. For the interior points of D, replace the gradient in c with the central difference in the x and y directions,

$$\Big(\frac{u(x_i + 0.2, y_i) - u(x_i - 0.2, y_i}{2\Delta x}, \frac{u(x_i, y_i + 0.2) - u(x_i, y_i - 0.2)}{2\Delta y}\Big).$$

Plot the vector field.

e. Compare the output in c against the output in d. Compute the mean normed error.

3. The function in this exercise arises in aerospace engineering. Consider the following function $f : \mathbb{C} \to \mathbb{C}$, $f(z) = z + 1/z$ defined on the complex plane, $z = x + \mathrm{i}y$. The inverse of f is given by

$$g(z) = \frac{1}{2}(z + \sqrt{z^2 - 4}).$$

Further, we can think of g as a function from $\mathbb{R}^2$ to $\mathbb{R}^2$,

$$g(x, y) = \frac{1}{2}(x + \mathrm{i}y + \sqrt{x^2 - y^2 + 2\mathrm{i}xy - 4}) = (Re(g(x, y)), Im(g(x, y))),$$

the real and imaginary parts of g. In fluid dynamics, it is important to know the real part of the gradient of g.

a. Write a program that approximates the gradient of g with central differences.

b. Extend the program in part a by taking points (x_0, y_0) in the domain of g and approximating the real part of the gradient at the location.

4.2 Finite Difference Method, Explicit or Forward Euler

In this section, we introduce finite difference method (FDM), one of basic techniques used to approximate the solution to a partial differential equation, PDE. This classical technique became more popular with the development of computers. The theory to support it dates from the 1950-1970, [Loustau (2016)]. Our treatment of the finite difference method given here and in the following sections is by no means exhaustive. It will serve to demonstrate that finite differences are useful. We begin our development with the particular form of FDM referred to as *explicit FDM* or *forward Euler FDM.*

Consider the following common setting. Suppose an unknown function is defined on an interval $[a, b]$ of the x-axis and a time interval $[0, T]$, $u = u(x, t)$, $u : [a, b] \times [0, T] \to \mathbb{R}$ and satisfies the following differential equation

$$\frac{\partial u}{\partial t} = \alpha \frac{\partial^2 u}{\partial x^2}. \tag{4.2.1}$$

This equation is called the *one dimensional heat equation*, as it can represent the diffusion of heat across a thin rod. However, this equation arises in settings, which have nothing to do with heat flow. Indeed, it represents a diffusion process in any context. For now, we focus on the usual heat flow setting. In this context, the values $u = u(x, t)$ represent temperatures at the location x and time t.

Suppose we partition the intervals $[a, b]$ and $[0, T]$ in a uniform manner, $a = x_0 < x_1 < \ldots < x_n = b$ and $0 = t_0 < t_1 < \ldots < t_N, \Delta x = x_i - x_{i-1}$ and $\Delta t = t_n - t_{n-1}$. We can recast (4.2.1) in finite difference format using the forward difference for time and the second central difference for space.

$$\frac{u(x_i, t_{n+1}) - u(x_i, t_n)}{\Delta t} = \alpha \frac{u(x_{i+1}, t_n) - 2u(x_i, t_n) + u(x_{i-1}, t_n)}{\Delta x^2} + \epsilon,$$

where ϵ is the cutoff error inherent in using finite differences for the derivatives. If we set $\lambda = \alpha \Delta t / \Delta x^2$, then we get the following expression,

$$u(x_i, t_{n+1}) = \lambda u(x_{i+1}, t_n) + (1 - 2\lambda) u(x_i, t_n) + \lambda u(x_{i-1}, t_n) + \epsilon.$$

For the heat context, this latter expression may be phrased as, the temperature at location x_i and the next time t_{n+1} is a linear combination of the temperatures at the current time t_n and location x_i together with the neighbors plus a small error. In other terms, we have a iterative process that infers temperatures in succeeding times from the current temperatures.

If we suppress the error term, then we should change the notation slightly to reflect that the numbers are not exact, but finite difference approximations. In this context, it is standard to write u_i^n for the value $u(x_i, t_n)$ computed with finite differences. With the new notation, we write

$$u_i^{n+1} = \lambda u_{i+1}^n + (1 - 2\lambda) u_i^n + \lambda u_{i-1}^n. \tag{4.2.2}$$

In the terminology of the topic, this rendering or choice of finite differences is referred to as *FTCS*, for forward time, central space.

It is important that the problem has unique solution. In the literature this is referred to setting a *well posed problem.* We generally assume that the underlying physics is deterministic. The requirement that the problem have a unique solution is consistent with this assumption. If we look at (4.2.1) we realize that any multiple of a solution u is also a solution. We can pin this down by designating the temperature at the boundary. In our finite difference setting, this is equivalent to declaring prior knowledge of the temperatures at the boundary. In particular, we must designate u_0^n and u_{k+1}^n for all n in order to have a well posed problem.

At this point, we develop the FDM in matrix notation. We already mentioned that each u_i^{n+1} is a linear combination of the entries in u_i^n, (4.2.2). We write u^n to denote the nth time state, the column vector whose entries are u_i^n. Now we have a linear relation between states, $u^{n+1} = Au^n$. We expand as follows

$$\begin{pmatrix} 1-2\lambda & \lambda & 0 & \dots & 0 & 0 & 0 \\ \lambda & 1-2\lambda & \lambda & \dots & 0 & 0 & 0 \\ 0 & \lambda & 1-2\lambda & \dots & 0 & 0 & 0 \\ \dots & \dots & \dots & \dots & \dots & \dots & \dots & \dots \\ 0 & 0 & 0 & \dots & 1-2\lambda & \lambda & 0 \\ 0 & 0 & 0 & \dots & \lambda & 1-2\lambda & \lambda \\ 0 & 0 & 0 & \dots & 0 & \lambda & 1-2\lambda \end{pmatrix} \begin{pmatrix} u_0^n \\ u_1^n \\ u_2^n \\ \dots \\ u_{k-1}^n \\ u_k^n \\ u_{k+1}^n \end{pmatrix} = \begin{pmatrix} u_0^{n+1} \\ u_1^{n+1} \\ u_2^{n+1} \\ \dots \\ u_{k-1}^{n+1} \\ u_k^{n+1} \\ u_{k+1}^{n+1} \end{pmatrix}. \tag{4.2.3}$$

The matrix in (4.2.3) is nonsingular. It is symmetric and the eigenvalues are known and nonzero for any k and any λ. (See [Loustau (2016)].) Therefore, given an initial state u^0, each successive state is completely determined, $u^n = A^n u^0$. Therefore, the 1D heat equation is inherently well posed as a FDM problem. But just as noted, any multiple of this matrix will also transform each state to the next and still be a discrete form of Equation (4.2.1). We remove this ambiguity by designating boundary values at $a = x_0$ and $b = x_{k+1}$. Generally, this data is known to us. In this case, we set values that may or may not be time dependent. For instance, if the boundary values at time t_{n+1} are μ_0^{n+1} and μ_{k+1}^{n+1}, then the linear relation (4.2.3) is modified as

$$\begin{pmatrix} 1 & 0 & 0 & \dots & 0 & 0 & 0 \\ \lambda & 1-2\lambda & \lambda & \dots & 0 & 0 & 0 \\ 0 & \lambda & 1-2\lambda & \dots & 0 & 0 & 0 \\ \dots & \dots & \dots & \dots & \dots & \dots & \dots & \dots \\ 0 & 0 & 0 & \dots & 1-2\lambda & \lambda & 0 \\ 0 & 0 & 0 & \dots & \lambda & 1-2\lambda & \lambda \\ 0 & 0 & 0 & \dots & 0 & 0 & 1 \end{pmatrix} \begin{pmatrix} \mu_0^{n+1} \\ u_1^n \\ u_2^n \\ \dots \\ u_{k-1}^n \\ u_k^n \\ \mu_{k+1}^{n+1} \end{pmatrix} = \begin{pmatrix} \mu_0^{n+1} \\ u_1^{n+1} \\ u_2^{n+1} \\ \dots \\ u_{k-1}^{n+1} \\ u_k^{n+1} \\ \mu_{k+1}^{n+1} \end{pmatrix}. \tag{4.2.4}$$

The transformation between the format (4.2.3) and (4.2.4) can be mathematically justified. See for instance [Loustau (2016)]. Of course, in order to use (4.2.4) we must have an initial state, u_i^0 for all locations x_i.

It is most important to realize that equation (4.2.1) defines a linear process that transforms the temperatures at one time state to the next. In the discrete form, we realize the linear process by the matrix in (4.2.4).

Consider the following example. Suppose there is a thin rod which is insulated along its length. Suppose that the temperature is initially zero everywhere, and that the left end is suddenly heated and kept at 20 degrees. Finally we set $\alpha = 1/2$. In notation, we have set the spatial interval is $[-5, 5]$, $\Delta x = 0.1$, $\Delta t = 0.01$ and $\lambda = 1/2$. For the initial setting take $u(x, 0) = 0$ and for the boundary values take $u(-5, t) = 20$ and $u(5, t) = 0$. Using (4.2.4), we solve for approximate values of u along the interval. The following plot (Figure 4.2.1) shows the local temperatures after 10 time steps.

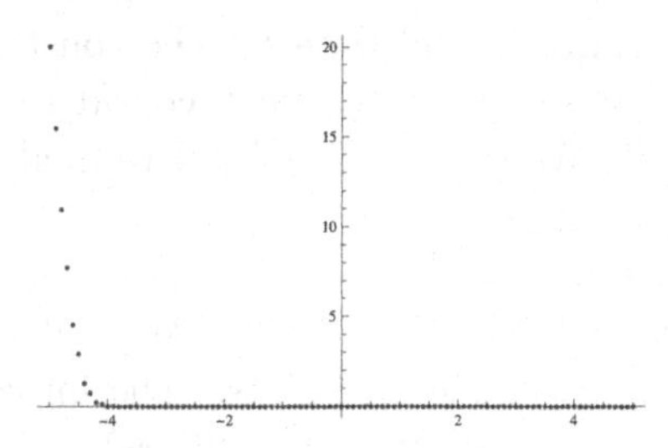

Figure 4.2.1: The temperature disribution after 10 iterations

This is the basic idea behind the finite difference method for solving differential equations. The explicit or forward Euler method and other related techniques are remarkably successful. In Section 4, we introduce two more FDM techniques. On the one hand any differential equation may be rendered in FDM form. However, all these techniques have their limitations. Compare the output from Exercises 1 and 2 or 5 and 6 below. The change in λ causes dramatic changes in the output. Indeed, the output of Exercise 2 is impossible. The difficulty arises because of computational error. In the next section, we develop a technique that demonstrates how λ is related to the error.

Problem 4 describes an application from cancer therapy. The specific context is the delivery of chemotherapy across a cell membrane.

Exercises:

1. Execute the example from the text using forward Euler method. Take $\alpha = 1/2$, the spatial interval as $[-5, 5]$, $\Delta x = 0.1$, $\Delta t = 0.01$. For the initial state take $u(x_i, 0) = 0$ for every i and for the boundary values take each $u(-5, t_n) = 20$ and $u(5, t_n) = 0$. Plot the temperatures at $t = 0.03, 0.06, 0.08, 0.1$.

2. Redo Problem 1 with $\alpha = 2$. Notice that the results are not well behaved. The problem here is that the forward Euler method is not stable then $\lambda > 0.5$. Stability of transient or time dependent processes is covered in following section.

3. Redo Problem 1 with $\alpha = 2$ and $\Delta t = 0.001$.

4. Diffusion of a chemical through a permeable membrane is sometimes modeled by $\partial u/\partial t = \alpha \partial^2 u/\partial x^2$, where $u(t, x)$ represents the concentration of the chemical at location x and time t. The constant α is determined by the membrane. This setting arises when considering the absorption of a substance into a cell. We want to use FDM to model this diffusion and determine when stasis occurs.

Consider a 4 point model, x_0 and x_1 outside the cell and x_2, x_3 inside the cell. Set $\alpha = 0.05$, $\Delta x = 0.25$, $\Delta t = 0.01$. Set up a FDM model with x_0 and x_3 as boundary points. The initial condition will be $u(0, x_0) = u(0, x_1) = 0.075$ and $u(0, x_2) = u(0, x_3) = 0.025$. We further assume that the concentration is constant at the boundary points. We want to determine t and $u(t, x_2)$ when stasis has been reached. For our purposes we define stasis as $u(t_{n+1}, x_2) - u(t_n, x_2) < 0.0001$.

5. Consider the *first order wave equation*. Given $u = u(t, x)$,

$$\frac{\partial u}{\partial t} = \alpha \frac{\partial u}{\partial x}.$$

a. Using forward differences, render this PDE in finite difference format. In this case, it is usual to write $C = \alpha \Delta t/\Delta x$. C is called the *Courant number*. Take $\alpha = -1/2$.

b. Execute FDM for the following setting.

$\alpha = -300$, Region: $[0, 300]$, Time interval $[0, 5]$,

Initial Conditions:

$u(0, x) = 0$, $0 \leq x \leq 50$;

$u(0, x) = 100 \sin\left[\pi \frac{(x-50)}{60}\right]$, $50 \leq x \leq 110$;

$u(0, x) = 0$, $110 \leq x \leq 300$.
Boundary Values: $u(t, 0) = 0$ and $u(t, 300) = 0$,
$\Delta x = 5$; $\Delta t = 0.0015$.
Use B-splines to display the output at time states 1, 3, 5.

6. Redo Problem 5 with $\alpha = 300$.

4.3 Neumann Stability Analysis

In the last section we developed a very general technique that seemed to allow us to resolve any second order PDE. Then in the exercises, we encountered problems. Further, there did not seem to be an apparent resolution. Certainly, if forward Euler FDM is going to be an important technique to approximate the solution to a PDE, we need to be able to predict when the calculations will yield reasonable results. And a step further, we want to know if the data we compute approximates the values of the actual solution. *Neumann stability analysis* resolves the first equation. We have more to say on the second later in the section.

Definition 4.3.1. An FDM realization is *Neumann stable* provided there is a positive integer n_0 so that the time state vectors satisfy $\|u^{n+1}\| \leq \|u^n\|$, for every $n \geq n_0$.

The idea here is that if the state vectors are non-increasing in norm, then the wild fluctuations in output that we experienced in the prior exercises cannot occur. In addition, there is a means to verify when the condition holds. As this technique uses the discrete Fourier interpolation we begin there.

We start with a uniform partition of an interval and a function f defined on the interval. With a simple variable transformation, we may suppose that the interval is $[-\pi, \pi]$. In particular, we have values $f(x_i)$ for $-\pi = x_0 < x_1 < ... < x_{N-1} < \pi$. Notice that the functions on this finite set are really just N-tuples $(f(x_0), f(x_1), ..., f(x_{N-1}))$. Hence, we may associate the set of all functions defined on the partition to $\mathbb{C}^N$. From results in linear algebra, we have a linear form σ on the space,

$$\sigma(f, g) = \sum_{i=0}^{N-1} f(x_i)\overline{g(x_i)}. \tag{4.3.1}$$

In Chapter 2, we defined a norm on a vector space. However, the function σ determines something weaker than a norm on the space of functions. The following definition modifies Definition 2.2.2.

Definition 4.3.2. Let V be a real or complex vector space. A *semi-norm* on V is a function $\|.\|$ taking values in $\mathbb{R}$ such that

(1) for any nonzero vector, $\|v\| \geq 0$,
(2) for any scalar α, $\|\alpha v\| = |\alpha|\|v\|$,
(3) for u and v in V, $\|u + v\| \leq \|u\| + \|v\|$.

Note that σ is not a norm on the space of functions defined on $[-\pi, \pi]$ as it is possible for $f(x_i) = 0$ while $f \neq 0$. The following theorem is a standard linear algebra result.

Theorem 4.3.1. *The linear form σ defines an Hermitian form on $\mathbb{C}^n$. Further, $\|f\|_\sigma = (\sigma(f, f))^{1/2}$ defines a semi-norm on $\mathbb{C}^n$.*

Proof. See [Loustau and Dillon (1993)]. □

Notice that we do not include $x_N = N\Delta x = \pi$. We cannot include this value because we do not want to assume or deny $f(\pi) = f(-\pi)$. In standard terminology, we do not assume that we have the function space of *periodic* functions.

We now define the discrete Fourier interpolation.

Definition 4.3.3. Let f be a complex valued function defined on the set $\{x_i : x_i = ih, i = 0, 1, ..., N - 1\}$. The *discrete Fourier interpolation* of f is given by

$$\hat{f}(x) = \sum_{k=-M_0}^{M_1 - 1} c_k e^{\mathbf{i}kx}, \tag{4.3.2}$$

where each $c_k = \sigma(f, e^{\mathbf{i}kx})/N$, and $M_0 = M_1 = N/2$ for N even, $M_0 = M_1 = (N - 1)/2$ for N odd. The c_k are called the *(discrete) Fourier coefficients* for f.

The expression in (4.3.2) is an interpolation of f as a trigonometric polynomial. In particular, $e^{\mathbf{i}kx} = \cos(kx) + \mathbf{i}\sin(kx)$. Hence, by the multiple angle identities, $\cos(kx)$ and $\sin(kx)$ may be written in terms of the powers of the cosine and sine. Therefore, the complex exponential is a polynomial in sine and cosine.

To put this into context, if we were to start with an absolutely integrable function f and form the Fourier transform and subsequently the reverse transform, then the result would be a function equal to f almost everywhere. In particular, the discrete Fourier interpolation is the inverse Fourier transform provided we replace the integration by the numerical integral using the trapezoid form. (See Section 5.1.)

The following theorems state the basic properties of the discrete Fourier interpolation. We do not prove all, as that would take us to far from our focus. Indeed, we restrict attention to the properties that are critical for FDM stability.

Lemma 4.3.1. *For any integers j and k,*

$$\sigma(e^{ijx}, e^{ikx}) = N,$$

if $(j-k)/N$ is an integer and

$$\sigma(e^{ijx}, e^{ikx}) = 0,$$

otherwise.

Proof. In any case, setting $h = 2\pi/N$,

$$\sigma(e^{\mathrm{i}jx}, e^{\mathrm{i}kx}) = \sum_{i=1}^{N-1} e^{\mathrm{i}jx_i} e^{-\mathrm{i}kx_i} = \sum_{i=0}^{N-1} e^{\mathrm{i}(j-k)ih} = \sum_{i=0}^{N-1} (e^{\mathrm{i}(j-k)h})^i.$$

If $(j-k)/N$ is an integer, then since $h = 2\pi/N$ we have

$$e^{\mathrm{i}(j-k)h} = e^{\mathrm{i}(j-k)2\pi/N} = (e^{2\pi\mathrm{i}})^{(j-k)/N} = 1^{(j-k)/N} = 1,$$

and

$$\sum_{i=0}^{N-1} (e^{\mathrm{i}(j-k)h})^i = N.$$

On the other hand, if $(j-k)/N$ is not an integer, then $e^{\mathrm{i}(j-k)h} = (e^{\mathrm{i}2\pi})^{(j-k)/N} = \zeta$, an Nth root of unity. We apply the usual expression for the partial geometric sum to get

$$\sum_{i=0}^{N-1} (e^{\mathrm{i}(j-k)h})^i = \sum_{i=0}^{N-1} \zeta^i = \frac{1-\zeta^N}{1-\zeta} = 0.$$

□

Theorem 4.3.2. *If $f(x_j) = \sum_{k=-M_0}^{M_1-1} d_k e^{ikx_j}$ for each $j = 0, 1, ..., N-1$, then each $d_k = c_k$ as given in Definition 4.3.2. Conversely, for each j, $\sum_{k=-M_0}^{M_1-1} c_k e^{ikx_j} = f(x_j)$.*

Proof. See [Loustau (2016)] or [Thomas (1999)] . □

Our next result and its corollary provide the mathematical foundation for Neumann stability analysis. The result identifies the norm of f as an N-tuple and the norm of $\hat{f}$ defined by σ. This is the discrete form of a well known theorem.

Theorem 4.3.3. *(Parceval) For f defined on the given partition, $f = (f(x_0), ..., f(x_{N-1}))$, then $\|f\| = \|\hat{f}\|_\sigma$, where $\|\hat{f}\|_\sigma$ is the semi-norm determined by σ, $(\sigma(\hat{f}, \hat{f}))^{1/2}$.*

Proof. As a complex N-tuple, the norm squared of f is

$$\|f\|^2 = \sum_{i=0}^{N-1} |f(x_i)|^2 = \sum_{i=0}^{N-1} f(x_i)\overline{f(x_i)} = \sigma(\hat{f}, \hat{f})$$

by the definition of σ. The result follows as $\sqrt{\sigma(\hat{f}, \hat{f})} = \|\hat{f}\|_\sigma$. □

Next, suppose that u^n is the state vector of an FDM process and suppose that u^n has discrete Fourier coefficients c_i^n, then Theorem 4.3.2 may be restated as follows.

Corollary 4.3.1. *An FDM realization is Neumann stable provided for each n and i, $|c_i^n| \geq |c_i^{n+1}|$.*

Proof. From the basic properties of the discrete Fourier interpolation and Theorem 2.3.2,

$$\|u^{n+1}\|^2 = \|\hat{u}^{n+1}\|_\sigma^2 = \sigma(\sum_{i=M_0}^{M_1-1} c_i^{n+1} e^{\mathrm{i}ix}, \sum_{j=M_0}^{M_1-1} c_j^{n+1} e^{\mathrm{i}jx})$$

$$= \sum_{i,j=M_0}^{M_1-1} c_i^{n+1}\overline{c_j^{n+1}}\sigma(e^{\mathrm{i}ix}, e^{\mathrm{i}jx}) = N \sum_{i=M_0}^{M_1-1} |c_i^{n+1}|^2$$

since $\sigma(e^{\mathrm{i}ix}, e^{\mathrm{i}jx}) = 0$ if $i \neq j$ and $\sigma(e^{\mathrm{i}jx}, e^{\mathrm{i}jx}) = N$. This last statement follows from Lemma 4.3.1. By hypothesis,

$$N \sum_{i=M_0}^{M_1-1} |c_i^{n+1}|^2 \leq N \sum_{i=M_0}^{M_1-1} |c_i^n|^2 = \|\hat{u}^n\|_\sigma^2 = \|u^n\|^2.$$

The result follows. □

We now return to the 1-D heat equation and the explicit formulation as derived from (4.2.2).

$$u_i^{n+1} = \lambda u_{i+1}^n + (1-2\lambda)u_i^n + \lambda u_{i-1}^n$$

If we write this equation as the discrete Fourier interpolation we get

$$\sum_{k=M_0}^{M_1-1} c_k^{n+1} e^{\mathbf{i}kx_i}$$

$$= \lambda \sum_{k=M_0}^{M_1-1} c_k^n e^{\mathbf{i}kx_{i+1}} + (1-2\lambda) \sum_{k=M_0}^{M_1-1} c_k^n e^{\mathbf{i}kx_i} + \lambda \sum_{k=M_0}^{M_1-1} c_k^n e^{\mathbf{i}kx_{i-1}}$$

$$= \sum_{k=M_0}^{M_1-1} (\lambda c_k^n e^{\mathbf{i}k\Delta x} + (1-2\lambda)c_k^n + \lambda c_k^n e^{-\mathbf{i}k\Delta x}) e^{\mathbf{i}ki\Delta x}$$

or

$$\sum_{k=M_0}^{M_1-1} c_k^{n+1} e^{\mathbf{i}ki\Delta x} = \sum_{k=M_0}^{M_1-1} (\lambda c_k^n e^{\mathbf{i}k\Delta x} + (1-2\lambda)c_k^n + \lambda c_k^n e^{-\mathbf{i}k\Delta x}) e^{\mathbf{i}ki\Delta x}.$$

By equating corresponding Fourier coefficients (Theorem 2.3.1)

$$c_j^{n+1} = \lambda c_j^n e^{\mathbf{i}j\Delta x} + (1-2\lambda)c_j^n + \lambda c_j^n e^{-\mathbf{i}j\Delta x}.$$

Now divide through by c_j^n and take the absolute value,

$$\frac{|c_j^{n+1}|}{|c_j^n|} = |\lambda e^{\mathbf{i}j\Delta x} + (1-2\lambda) + \lambda e^{-\mathbf{i}j\Delta x}|.$$

Expressing the right hand side in terms of cosine and sine yields,

$$\frac{|c_j^{n+1}|}{|c_j^n|} = |\lambda(\cos(j\Delta x) + \mathbf{i}\sin(j\Delta x))$$

$$+(1-2\lambda) + \lambda(\cos(j\Delta x) - \mathbf{i}\sin(j\Delta x))|,$$

or

$$\frac{|c_j^{n+1}|}{|c_j^n|} = |2\lambda(\cos(j\Delta x) + (1-2\lambda)| = |1 + 2\lambda(\cos(j\Delta x) - 1)|.$$

Hence, we want to know when the fraction on the left hand side is less than or equal to 1. Equivalently, $-1 \le 1 - 2\lambda(1 - \cos(j\Delta x)) \le 1$, or $-2 \le -2\lambda(1 - \cos(j\Delta x)) \le 0$. This reduces to $1 \ge \lambda(1 - \cos(j\Delta x))$. But the maximal value of $1 - \cos(\varphi)$ is 2. Hence, $\lambda \le 1/2$. Appealing to Theorem 4.3.3, we have proved the following theorem.

Theorem 4.3.4. *The FDM rendering of the 1D heat equation is von Neumann stable if and only if* $\lambda \le 0.5$.

For the proof of Theorem 4.3.4, we were able to take advantage of some simple properties of the trigonometry functions. Alternatively, we could have applied basic calculus max/min procedures to determine values of λ that satisfy $1 \geq -2\lambda(1 - \cos(j\Delta x))$.

We see now that stability depends on the balance between Δx and Δt. Generally, if you decrease Δx then you must decrease Δt so that it accommodates for the square of Δx.

But stability is only one issue. We also must consider the error, $e_n(x_i) = u(t_n, x_i) - u_i^n$. We can see that this is related to stability via the matrix A. Denoting the ith row of A by $A_{(i)}$, we calculate.

$$e_n(x_i) = u(t_n, x_i) - u_i^n = u(t_n, x_i) - A_{(i)} u_i^{n-1}$$

$$= u(t_n, x_i) - A_{(i)} u(t_{n-1}, x_i) + A_{(i)} u(t_{n-1}, x_i) - A_{(i)} u_i^{n-1}$$

$$= u(t_n, x_i) - A_{(i)} u(t_{n-1}, x_i) + A_{(i)}[u(t_{n-1}, x_i) - u_i^{n-1}]$$

$$= u(t_n, x_i) - A_{(i)} u(t_{n-1}, x_i) + A_{(i)} e_{n-1}(x_i) = c_n(x_i) + A_{(i)} e_{n-1}(x_i),$$

where $c_n(x_i) = u(t_n, x_i) - A_{(i)} u(t_{n-1}, x_i)$. If we continue the steps regressing back to the initial stage, we may write,

$$e_n(x) = c_n(x) + A e_{n-1}(x)$$

$$= ... = c_n(x) + A c_{n-1}(x) + A^2 c_{n-2}(x) + ... + A^{n-1} c_1(x) + A^n e_0(x)$$

$$= c_n(x) + A c_{n-1}(x) + A^2 c_{n-2}(x) + ... + A^{n-1} c_1(x),$$

since the initial error is zero. We will not present the proof that $e_n \to 0$ as $n \to \infty$. From the prior equation we see that $e_n \to 0$ provided c_n is bounded and the powers of A are bounded in operator norm. For the details see [Loustau (2016)].

In this last equation there are two terms, the powers of A and the vectors c_n. We expect that the FDM is stable when the sequence $A^n \to 0$ in operator norm. This is true for FDM applied to the 1-D heat equation. Further, if u is C^2 then the sequence $c_n \to 0$ as Δx and $\Delta t \to 0$. To prove this, we would apply the Taylor expansions used to derive the finite difference operators. Hence, it follows that $e_n \to 0$ and we know that the values computed by FDM are indeed good approximations of the actual values of u. For other equations the situation is not so nice. These questions are for a more advanced text.

Exercises:

1. Recall the first order wave equation, which arose in the exercises of Section 4.2. Given $u = u(t, x)$,

$$\frac{\partial u}{\partial t} = \alpha \frac{\partial u}{\partial x}$$

with FDM rendering, forward time and forward space (FTFS).

a. Implement FTFS for the first order wave for $\alpha = 1/2$. And also for $\alpha = -1/2$.

b. Execute Neumann stability analysis for $\alpha > 0$.

c. Execute Neumann stability analysis for $\alpha < 0$.

d. What is the difference between the case for α positive and α negative.

2. Forward time, central space provides an alternative FDM rendering of the order 1 wave equation. Resolve stability for this FDM procedure.

3. Consider the equation

$$\frac{\partial u}{\partial t} = \frac{\partial^2 u}{\partial x^2} + u.$$

a. Determine the explicit FDM rendering for this equation. What is the A matrix?

b. When is FTCS stable for this equation.

4.4 Finite Difference Method, Implicit and Crank Nicolson

In this section we look at two alternative FDM approaches for the 1D heat equation. The first uses backward time, central space, BTCS. With the notation of Section 4.2,

$$u_i^{n+1} - u_i^n = \lambda u_{i+1}^{n+1} - 2\lambda u_i^{n+1} + \lambda u_{i-1}^{n+1}.$$

After solving for u_i^n we have the following expression analogous to (4.2.2).

$$u_i^n = -\lambda u_{i+1}^{n+1} + (1 + 2\lambda) u_i^{n+1} - \lambda u_{i-1}^{n+1}. \tag{4.4.1}$$

As in Section 2, we use (4.4.1) to form a linear relation $Au^{n+1} = u^n$. Note that this relation is the reverse of 4.2.2, as A maps the $n + 1^{st}$ state to the nth state.

In matrix form, after setting boundary values, this yields $Bu^{n+1} = u^n$,

$$\begin{pmatrix} 1 & 0 & 0 & ... & 0 & 0 & 0 \\ -\lambda & 1+2\lambda & -\lambda & ... & 0 & 0 & 0 \\ 0 & -\lambda & 1+2\lambda & ... & 0 & 0 & 0 \\ ... & ... & ... & ... & ... & ... & ... & ... \\ 0 & 0 & 0 & ... & 1+2\lambda & -\lambda & 0 \\ 0 & 0 & 0 & ... & -\lambda & 1+2\lambda & -\lambda \\ 0 & 0 & 0 & ... & 0 & 0 & 1 \end{pmatrix} \begin{pmatrix} \mu_0^{n+1} \\ u_1^{n+1} \\ u_2^{n+1} \\ ... \\ u_{k-1}^{n+1} \\ u_k^{n+1} \\ \mu_{k+1}^{n+1} \end{pmatrix} = \begin{pmatrix} \mu_0^{n+1} \\ u_1^n \\ u_2^n \\ ... \\ u_{k-1}^n \\ u_k^n \\ \mu_{k+1}^{n+1} \end{pmatrix}. \tag{4.4.2}$$

The process is called *backward Euler* or *implicit* FDM. As in the explicit case, the matrix B is nonsingular. Indeed, there are formula for the eigenvalues of B and none of them are zero or near to zero. See [Loustau (2016)].

In this case, we know the nth state and want to solve for the $n+1^{st}$ state. Hence, either we solve (4.4.2) as a linear system or we compute the inverse of B and resolve $u^{n+1} = B^{-1}u^n$. Hence, $u^{n+1} = B^{-(n+1)}u^0$ provided the boundary values are time independent.

We know from experience that we should consider stability. Analogous to the explicit case, we resolve this question by recasting (4.4.1) in Fourier form.

$$\sum_{k=M_0}^{M_1-1} c_k^n e^{\mathbf{i}kx_i}$$

$$= -\lambda \sum_{k=M_0}^{M_1-1} c_k^{n+1} e^{\mathbf{i}kx_{i+1}} + (1+2\lambda) \sum_{k=M_0}^{M_1-1} c_k^{n+1} e^{\mathbf{i}kx_i} - \lambda \sum_{k=M_0}^{M_1-1} c_k^{n+1} e^{\mathbf{i}kx_{i-1}}$$

$$= \sum_{k=M_0}^{M_1-1} (-\lambda c_k^{n+1} e^{\mathbf{i}k\Delta x} + (1+2\lambda) c_k^{n+1} - \lambda c_k^{n+1} e^{-\mathbf{i}k\Delta x}) e^{\mathbf{i}ki\Delta x}$$

or

$$\sum_{k=M_0}^{M_1-1} c_k^n e^{\mathbf{i}ki\Delta x}$$

$$= \sum_{k=M_0}^{M_1-1} (-\lambda c_k^{n+1} e^{\mathbf{i}k\Delta x} + (1+2\lambda) c_k^{n+1} - \lambda c_k^{n+1} e^{-\mathbf{i}k\Delta x}) e^{\mathbf{i}ki\Delta x}.$$

By equating corresponding Fourier coefficients (Theorem 2.3.1)

$$c_j^n = \lambda c_j^{n+1} e^{\mathbf{i}j\Delta x} + (1 - 2\lambda) c_j^{n+1} + \lambda c_j^{n+1} e^{-\mathbf{i}j\Delta x}.$$

Divide through by c_j^{n+1} and take the absolute value,

$$\frac{|c_j^n|}{|c_j^{n+1}|} = |-\lambda e^{\mathbf{i}j\Delta x} + (1+2\lambda) - \lambda e^{-\mathbf{i}j\Delta x}|.$$

Expressing the right hand side in terms of cosine and sine yields

$$\frac{|c_j^n|}{|c_j^{n+1}|} = |-\lambda(\cos(j\Delta x) + \mathbf{i}\sin(j\Delta x))$$

$$+(1+2\lambda) - \lambda(\cos(j\Delta x) - \mathbf{i}\sin(j\Delta x))|$$

$$= |1 + 2\lambda(1 - \cos(j\Delta x))| \geq 1,$$

since $1 - \cos(j\Delta x) \geq 0$ and $\lambda > 0$. Therefore, $|u^{n+1}|/|u^n| \leq 1$ unconditionally. We have proved the following theorem.

Theorem 4.4.1. *Implicit FDM applied to the 1D heat equation is unconditionally stable.*

At this stage, it is natural to ask why we bother with explicit FDM when implicit is so well behaved. There are several reasons why implicit FDM is not sufficient. First, if the PDE is not linear, then explicit FDM is often the time stepping procedure of choice. (See [Loustau (2016)]). Secondly, implicit FDM often under estimates the result so badly that it is useless. There are instances in the literature where authors suggest executing a few steps with explicit FDM and then changing over once the data is well initialized. The third reason is the explicit and implicit combine to form the Crank Nicolson method. This is by far the most popular of the FDM time stepping procedures for the heat equation and related PDE.

To begin *Crank Nicolson*, we write the time step as the average of the implicit and explicit formulations,

$$u_i^{n+1} - u_i^n = \frac{1}{2}\left[\lambda u_{i+1}^{n+1} - 2\lambda u_i^{n+1} + \lambda u_{i-1}^{n+1}\right] + \frac{1}{2}\left[\lambda u_{i+1}^n - 2\lambda u_i^n + \lambda u_{i-1}^n\right].$$

Next, we introduce a ficticious intermediate time state, $u^{n+1/2}$,

$$(u_i^{n+1} - u_i^{n+1/2}) + (u_i^{n_1/2} - u_i^n)$$

$$= \frac{1}{2}\left[\lambda u_{i+1}^{n+1} - 2\lambda u_i^{n+1} + \lambda u_{i-1}^{n+1}\right] + \frac{1}{2}\left[\lambda u_{i+1}^n - 2\lambda u_i^n + \lambda u_{i-1}^n\right].$$

We separate this equation into a two step process,

$$(u_i^{n+1} - u_i^{n+1/2}) = \frac{1}{2}\left[\lambda u_{i+1}^{n+1} - 2\lambda u_i^{n+1} + \lambda u_{i-1}^{n+1}\right]$$

$$(u_i^{n+1/2} - u_i^n) = \frac{1}{2}\left[\lambda u_{i+1}^n - 2\lambda u_i^n + \lambda u_{i-1}^n\right].$$

The step $u^n \to u^{n+1/2}$ is explicit while the step $u^{n+1/2} \to n^{n+1}$ is implicit,

$$u_i^{n+1/2} = \frac{1}{2}\left[-\lambda u_{i+1}^{n+1} + 2(1+\lambda)u_i^{n+1} - \lambda u_{i-1}^{n+1}\right],$$

$$u_i^{n+1/2} = \frac{1}{2}\left[\lambda u_{i+1}^n - 2(1+\lambda)u_i^n + \lambda u_{i-1}^n\right].$$

The corresponding matrices are

$$\frac{1}{2}\begin{pmatrix} 1 & 0 & 0 & \dots & 0 & 0 & 0 \\ \lambda & -2(1+\lambda) & \lambda & \dots & 0 & 0 & 0 \\ 0 & \lambda & -2(1+\lambda) & \dots & 0 & 0 & 0 \\ \dots & \dots & \dots & \dots & \dots & \dots & \dots\ \dots \\ 0 & 0 & 0 & \dots & -2(1+\lambda) & \lambda & 0 \\ 0 & 0 & 0 & \dots & \lambda & -2(1+\lambda) & \lambda \\ 0 & 0 & 0 & \dots & 0 & 0 & 1 \end{pmatrix},$$

and

$$\frac{1}{2}\begin{pmatrix} 1 & 0 & 0 & \dots & 0 & 0 & 0 \\ -\lambda & 2(1+\lambda) & -\lambda & \dots & 0 & 0 & 0 \\ 0 & -\lambda & 2(1+\lambda) & \dots & 0 & 0 & 0 \\ \dots & \dots & \dots & \dots & \dots & \dots & \dots\ \dots \\ 0 & 0 & 0 & \dots & 2(1+\lambda) & -\lambda & 0 \\ 0 & 0 & 0 & \dots & -\lambda & 2(1+\lambda) & -\lambda \\ 0 & 0 & 0 & \dots & 0 & 0 & 1 \end{pmatrix}.$$

The first yields the $n + 1/2$ state from the nth. The inverse of the second maps the $n + 1/2$ state to the $n + 1^{st}$.

Taken together they are unconditionally stable.

Theorem 4.4.2. *Crank Nicolson FDM applied to the 1D heat equation is unconditionally stable.*

Proof. See Exercie 4. □

Exercises:

1. Do Exercise 1 of Section 4.2 using implicit FDM.

2. Do Exercise 1 of Section 4.2 using Crank Nicolson FDM.

3. Compare the three treatments for the setting in Exercise 1 of Section 4.2. Interpret the result in light of the statesmen that Crank Nicolson is the procedure of choice.

4. Prove Theorem 4.4.2.

a. Use the discrete Fourier interpolation to get expressions for

$$\frac{|c_k^{n+1}|}{|c_k^{n+1/2}|}, \quad \frac{|c_k^{n+1/2}|}{|c_k^n|}.$$

b. Multiply the two expression to get and expression for $|c_k^{n+1}|/|c_k^n|$.

5. Crank Nicolson for the first order wave equation.

a. Implement FTCS and BTCS for the first order wave equation.

b. Implement Crank Niicolson for the first order wave equation.

c. Do the Neumann stability analysis for this process.

d. The result of Part c was $|c_k^{n+1}|/|c_k^n| = 1$. Is such a procedure numerically robust? Why?

Chapter 5

Numerical Integration

Introduction

Numerical integration is one of the oldest numerical analysis topics. These techniques approximate integrals using only a few function values. Hence, they can be applied to observational data, cases where we may have little knowledge of the underlying function. Additionally, there are cases where we know the function, but the integral is intractable or worse, is a function with no closed form anti-derivative. For instance B-splines and Bezier curves are parametric cubics, $(x(t), y(t))$. By solving $x(t)$ for t and substituting into the $y(t)$, you can represent segments of the curve as a function graph $(x, f(x))$. Invariably, functions derived from a parametric cubic in this fashion rarely have closed form anti-derivatives. It is also the case that the integral that arises in arc length computations is rarely resolvable.

We begin the chapter with three numerical integration techniques. We encounter two in calculus, the trapezoid, midpoint method. They are used to develop the idea of the Riemann integral as the area under a curve. Simpson's rule can be expressed as a linear combination of the other two. We will see that both midpoint and trapezoid are exact for degree 1 polynomials. Simpson's method is exact for quadratics.

The quadrature procedures use polynomial interpolation to estimate the integral. These techniques proceed as follows. Given a function f, we use an interpolation procedure to determine a polynomial p. Next, we integrate p. This step is routine. As the interpolation procedure has an error estimate, then we can measure how well the integral of p approximates the integral of f.

Gaussian quadrature is the main event of this chapter. Generally speaking, this is considered the best of the numerical integration procedures. The underlying interpolation is Hermite. Since Hermite interpolation at $n+1$ locations is exact for polynomials for degree $2n+1$, then Gaussian quadrature is also exact for polynomials of degree $2n+1$.

There is an important technique that uses Gaussian quadrature to resolve integrals over a triangle. We develop this method in the last section. At this time, we also include weighted quadrature, in particular Chebyshev quadrature. Finally, we develop numerical integration for functions in parametric form.

Before proceeding we note that *Mathematica* has two integration commands, $\int_a^b f[x]dx$ is equivalent to $Integrate[f[x], \{x,a,b\}]$. In this case, *Mathematica* will compute the integral of f by identifying the anti-derivative g and evaluating $g(b) - g(a)$. If you use $NIntegrate[f[x], \{x,a,b\}]$, then the system executes Gaussian quadrature. Alternatively, if *Mathematica* fails to find an anti-derivative or the function is one of the functions known to not have an anti-derivative, then *Mathematica* will automatically execute Gaussian quadrature.

We make a final comment. The term *quadrature* refers to approximating an area by filling the region with quadrilaterals and then summing the areas of the simple figures. The trapezoid and midpoint methods are examples. However, the term survives today as a generic expression for numerical integration.

5.1 Trapezoid Method and Simpson's Rule

Perhaps the conceptually simplest of the numerical integration procedure is the trapezoid method. We begin with a function f defined on an interval $[a,b]$ that is bounded, real valued and integrable. Furthermore, suppose we have a partition of $[a,b]$, $a \le x_0 < x_1 < ... < x_n \le b$. If we know f or at least have the values $f(x_i)$, then we can easily approximate the integral of f over $[a,b]$. Indeed, if f is positive on $[a,b]$ and we join the points $(x_i, f(x_i))$ with line segments, then the area under the resulting polygon will approximate the area under f. Since the figure formed by connecting the four points $(x_{i-1},0)$, $(x_i,0)$, $(x_i,f(x_i))$ and $(x_{i-1},f(x_{i-1}))$ is a trapezoid, the area under the polygon as the sum of areas of trapezoids. (See Figure 5.2.1.)

Since the area of a trapezoid is height times the average of the bases, we write

$$\int_a^b f dx \approx \sum_{i=1}^{n} \frac{f(x_i)+f(x_{i-1})}{2}(x_i - x_{i-1}). \tag{5.1.1}$$

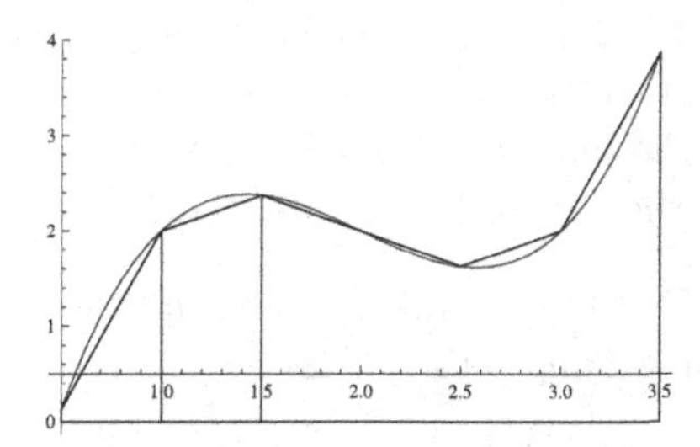

Figure 5.1.1: The temperature disribution after 10 iterations

In this case the height is measured along the x-axis and the bases are vertical.

Notice that if f is negative on [a, b], then (5.1.1) will evaluate negatively for each trapezoid associated to the integral of a negatively valued function. If $f(x_i) = 0$, then the corresponding trapezoid degenerates to a right triangle with area $(1/2)f(x_{i-1})(x_i - x_{i-1})$. Hence, (5.5.1) continues to hold. Finally, consider the case where f takes both positive and negative values along the partition. Without loss of generality $f(x_i) > 0$ and $f(x_{i-1}) < 0$. In this case, there is no trapezoid, rather the figure formed by the four points $(x_i, 0)$, $(x_{i-1}, 0)$, $(x_i, f(x_))$ and $(x_{i-1}, f(x_{i-1}))$ is the union of two right triangles that meet at a single point. One triangle is above the x-axis and the other is below. We set $\hat{x}$ to be the point where the line connecting $(x_1, f(x_i))$ and $(x_{i-1}, f(x_{i-1}))$ meets the x-axis and take $f(\hat{x}) = 0$. Taking $\hat{x}$ as a point in the partition, we have returned to the prior case. All together the right hand side of Equation 5.1.1 provides an approximation for the integral on the left hand side. Of course, this technique is only effective if the partition of $[a, b]$ is sensitive to changes in f.

Definition 5.1.1. Suppose f is defined on an interval $[a, b]$ with a partition given above, then approximation of the integral of f given by (5.1.1) is called the *trapezoid method.*

The trapezoid method is intuitively obvious, easy to implement and hence, a reasonable procedure. Later, we will learn that it is not as accurate as another less intuitive processes. In Exercise 4 we lead the reader to an error estimate for the trapezoid method.

Next, we look at an alternative to the trapezoid method. Notice that the area of the trapezoid returns the actual value of the integral when the function is linear. But most functions that we consider have curved graphs, not polygonal graphs. Simpson's Rule is characterized by the fact that it returns the actual integral for quadratics. In other words, you would use it when the graph of the function is more or less a parabola. First we state Simpson's rule and then we will verify the claim.

Definition 5.1.2. For the interval $[a-h, a+h]$, the estimated integral for f in the interval using *Simpson's rule* is

$$\int_{a-h}^{a+h} f(x)dx \approx 2h\big[\frac{1}{6}f(a+h) + \frac{2}{3}f(a) + \frac{1}{6}f(a-h)\big]. \tag{5.1.2}$$

If we have an interval $[a, b]$ together with a partition, $a = x_0 < x_1 < ... < x_n = b$. Then Simpson's rule states that

$$\int_a^b f(x)dx \approx \sum_{i=1}^{n} \big[\frac{1}{6}f(x_i) + \frac{2}{3}f(\frac{x_i + x_{i-1}}{2}) + \frac{1}{6}f(x_{i-1})\big](x_i - x_{i-1}). \tag{5.1.3}$$

In the second statement we merely recast the technique in the same format as the trapezoid method. Notice that $2h$ becomes $(x_i - x_{i-1})$.

If $f(x) = (x-a)^2$, then Simpson's method yields

$$2h\big[\frac{1}{6}f(a+h) + \frac{2}{3}f(a) + \frac{1}{6}f(a-h)\big] = 2h\big[\frac{1}{6}h^2 + \frac{1}{6}h^2\big] = \frac{2}{3}h^3,$$

which is exactly the integral of f. Therefore, Simpson's Rule is exact in the case of the quadratic polynomial centered at the interval midpoint. This issue is fully resolved in Exercise 3.

Consider the $f(x) = x^3$, so that $\int_0^1 f(x)dx = 1/4$. Now take $x_0 = 0$, $x_1 = 1/4$, $x_2 = 1/2$, $x_3 = 3/4$, $x_4 = 1$. Whereas, the trapezoid approximation is 0.265625, Simpson's rule evaluates it at 0.25 to 5 decimal places. Hence, this method also yields good results for the cubic.

As mentioned, one of the important applications for numerical integration is computing arc length. In particular, if $f(x) = y$ for x in the interval $[a, b]$, then the length of the graph of f is given by

$$\int_a^b [1 + f'(x)^2]^{1/2}dx. \tag{5.1.4}$$

But if the curve is given parametrically, $\gamma(t) = (\gamma_1(t), \gamma_2(t))$, $t \in [a, b]$, then the arc length is known directly from the parametric formulation as

$$\int_a^b [\gamma_1'(t)^2 + \gamma_2'(t)^2]^{1/2}dt. \tag{5.1.5}$$

Again, the presence of the square root often renders the integral intractable. As a result, we are left with numerical methods as the only way forward.

It is not difficult to resolve the integral in (5.1.4). The integral in (5.1.5) is different. In this case, we have the following problem. Given α on the x-axis, we must find β so that (α, β) lies on γ. We discussed it briefly in the introduction. We return to this question in Section 4.

Exercises:

1. Let $f(x) = xe^{-x} - 1$ and set the interval to $[1, 4]$.

a. Use integration by parts to compute the integral of f on the given interval. Use *Mathematica* to evaluate the exponentials.

b. Use the Integrate command in *Mathematica.* Compare the result with the result in (a).

2. Let $f(x) = xe^{-x} - 1$ and set the interval to $[1, 4]$. Consider the partition $1 < 1.5 < 2 < 2.5 < 3 < 3.5 < 4$.

a. Compute the integral of f using the trapezoid method for the given partition.

b. Compute the integral of f using Simpson's rule for the given partition. Compare these results with those in 1.

3. Prove that Simpson's rule is exact for quadratics. Hint: For $f(x) = ax^2 + bx + c$, prove that we can write f in the form $f(x) = \alpha(x - \beta)^2 + \gamma$.

4. Derive the following estimate for the trapezoid method error. Suppose we have a function f defined on $[a, b]$ with partition $a = x_0 < x_1 < ... < x_n = b$. Let T denote the trapezoid method approximation for the integral. Suppose that f is $n + 2$ times continuously differential on $[a, b]$.

a. Let p be the polynomial interpolation of f associated to the given partition. Go to Section 3.1 to find an expression for the error $e(x) = f(x) - p(x)$.

b. Find the expression for the upper bound of $|e(x)|$. Use this expression to derive an estimate for $| \int_a^b f - \int_a^b p|$.

c. Compute the difference $\int_a^b p - T$.

d. Combine the prior steps to derive an estimate for $\int_a^b f - T$.

5.2 Midpoint Method

The midpoint is remarkable both because of its simplicity and its accuracy. As before, we begin with a function f defined on an interval $[a, b]$ and a partition $a \leq x_0 < x_1 < ... < x_n \leq b$. For the sub-interval $[x_{i-1}, x_i]$, let α denote the midpoint, $(x_{i-1} + x_i)/2$ of the interval and $2h$ the length. The interval now becomes $[\alpha + h, \alpha + h]$ and the trapezoid approximation for the integral is $2h(f(\alpha + h) + f(\alpha - h))/2 = h(f(\alpha + h) + f(\alpha - h))$. If we were to replace $[f(\alpha + h) + f(\alpha - h)]/2$ by $f(\alpha)$, then the method would be the midpoint rule.

Definition 5.2.1. For the interval $[a - h, a + h]$, the estimated integral for f in the interval using *midpoint rule* is

$$\int_{a-h}^{a+h} f(x)dx \approx 2hf(a). \tag{5.2.1}$$

If we have an interval $[a, b]$ together with a partition, $a \leq x_0 < x_1 < ... < x_n \leq b$, then midpoint rule states that

$$\int_a^b f(x)dx \approx \sum_{i=1}^{n} f(\alpha_i)(x_i - x_{i-1}), \tag{5.2.2}$$

where each $\alpha_i = [x_i + x_{i-1}]/2$.

Given a function f defined and integrable on a closed interval $[a, b]$, we write $s(f, a, b)$ for the Simpson's rule integral estimate, $m(f, a, b)$ for the midpoint and $t(f, a, b)$ for the trapezoid integral estimate. Then $s(f, a, b) = (2/3)t(f, a, b) + (4/3)m(fa, b)$.

The following theorem states that in important cases, the midpoint error is less than the trapezoid error.

Theorem 5.2.1. *Suppose the f is concave up or concave down of the interval $[a, b]$, the absolute numerical integration error for the midpoint method is less than or equal to the absolute error for the trapezoid method.*

Proof. The result follows from a simple argument using elementary geometry.

Suppose that the function is concave up in the interval. See Figure 3.1a. The error for the midpoint rule is the area between the horizontal line and the curve. The area to the left of the point A, denoted by α, is positive and the area on the right is negative. Denoting the error to the right as $-\beta$ for $\beta > 0$, we have the midpoint method error is $\alpha - \beta$.

If we include the tangent line to f at the midpoint (see Figure 5.2.1b), then triangles ABC and ADE are congruent (use side-angle-side). Let ζ be the common area of the two triangles, then $\beta = \zeta + \delta$ and $\alpha = \zeta - \gamma$. Hence, we may rewrite the midpoint error as $\alpha - \beta = \zeta - \gamma - (\zeta + \delta) = -(\gamma + \delta)$.

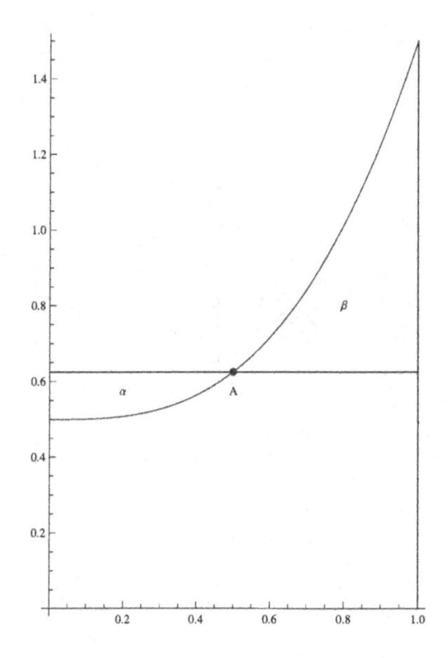

Figure 5.2.1a: Midpoint method

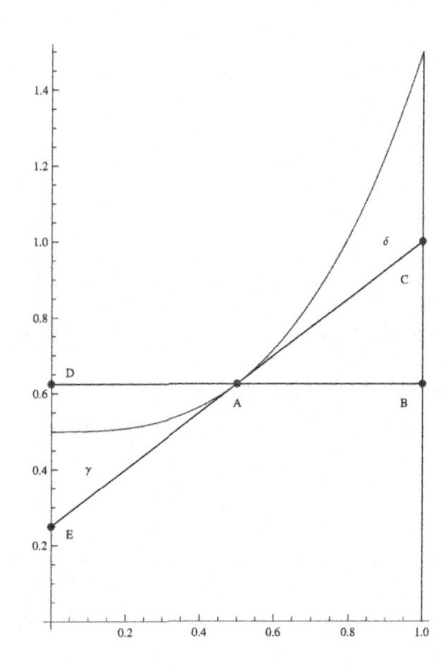

Figure 5.2.1b. Midpoint method with tangent at the midpoint

This last term is in fact the error if we had used the tangent line to compute the approximate integral. Taking absolute values, the absolute midpoint method error is $\gamma + \delta$.

Figure 5.2.2a shows the trapezoid method error, $\xi + \eta$, along with the tangent line at the interval midpoint. Next we construct line segments from M and N parallel to the tangent. (See Figure 5.2.2b.) We now have two parallelograms, $P_1 : EASM$ and $P_2 : ACNT$. Since the triangles MSQ and QNT are congruent (use angle-side-angle), then $\xi + \eta$ is equal to the part of P_1 and P_2 that lies above the curve. On the other hand the absolute midpoint method error is equal to the portion of the area of the two parallelograms that lies below the curve. Since the curve is concave up, then necessarily, $\xi + \eta > \gamma + \delta$, and we have proved the theorem. □

It is important to keep in mind that the midpoint method is easier to implement than trapezoid. And we can select the partition so that it is better. In the next section we will see that midpoint method is also the simplest case of Gaussian quadrature.

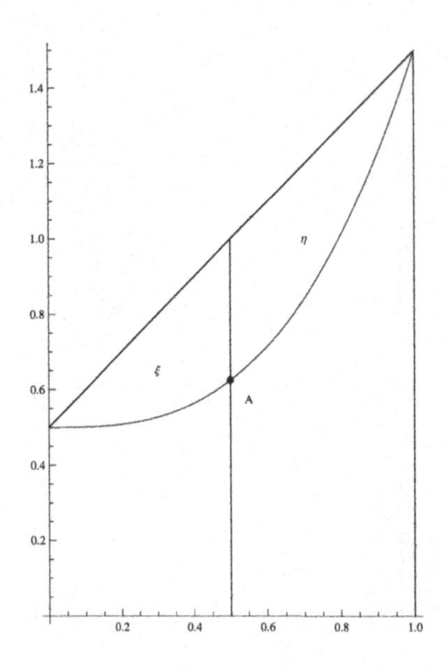

Figure 5.2.1c: The trapezoid method

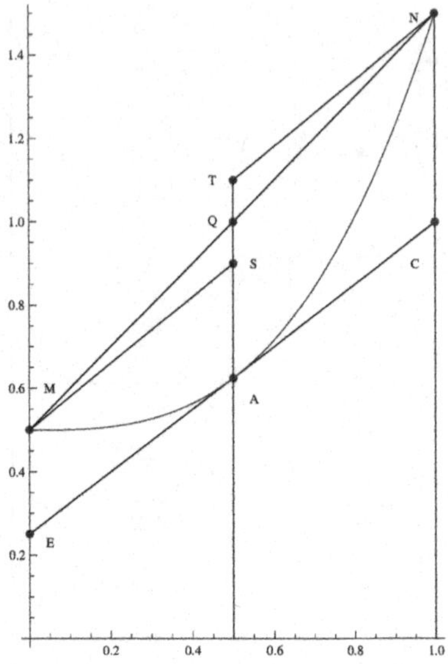

Figure 5.2.1d: With parallelograms

Exercises:

1. Let $f(x) = xe^{-x} - 1$ and set the interval to $[1, 4]$. Consider the partition $1 < 1.5 < 2 < 2.5 < 3 < 3.5 < 4$. Compute the integral of f using the midpoint rule. Compare these results with those of Exercises 1 and 2 of Section 5.1.

2. The following is a modified version of an often used numerical problem. In this exercise, we determine the heat generated by an exo-thermal reaction. Suppose two substances are placed in a kiln. The first substance is actively producing an exo-thermal reaction. The second substance is inert. Over a 25 minute period the temperature of the kiln goes from 86.2 to 126.5 degrees (Fahrenheit). At each minute the temperature of each substance is measured and the temperature difference is recorded. The results are in the following tables.

0	1	2	3	4	5	6	7	8
0.0	0.34	1.86	4.32	8.07	13.12	16.8	18.95	18.07

9	10	11	12	13	14	15	16
16.69	15.25	13.86	12.58	11.4	10.33	8.95	6.46

17	18	19	20	21	22	23	24	25
4.65	3.37	2.4	1.76	1.26	0.88	0.63	0.42	0.3

a. Plot the 26 points $(t, \delta(t))$. Using least squares, fit a quintic polynomial to the data and display both plots on the same axis. We denote this curve by p.

b. The temperature difference δ is caused by the reaction. The first t for which $\delta(t) > 0$ is the starting time of the reaction, denoted a. To find the end time of the reaction plot the points $(t, \log(\delta(t))$. You will notice that after a while the points seem to lie on a line. The value of t for which the plot begins to appear linear is denoted b and is the end time for the reaction.

c. Estimate the integral of δ using the midpoint rule applied to p. Use the partition that arises by subdividing the interval $[a, b]$ into 10 subintervals with length $h = (b - a)/10$.

d. Estimate the integral $\int_a^b \delta(t)dt$ using the trapezoid method. Use the same partition as in Part c.

The value calculated in c is called the *reaction heat.*

5.3 Gaussian Quadrature

The mathematics underlying Gaussian quadrature is polynomial interpolation. If f is a function with polynomial interpolation q and error e, then $f = q + e$ and it follows that $\int_a^b f = \int_a^b q + \int_a^b e$. Provided we know q then it is routine to calculate this integral. Hence, $\int_a^b f \approx \int_a^b q$, and we can use $\int_a^b e$ to estimate the error. The interpolation of choice is a special case of the Hermite interpolation.

We begin with some terminology. Recall that a positive definite inner product [Loustau and Dillon (1993)] is a real valued function σ of two variables defined on a vector space V so that for each u, v and w,

- $\sigma(u, v) = \sigma(v, u)$
- $\sigma(\alpha u + \beta v, w) = \alpha\sigma(u, w) + \beta\sigma(v, w)$
- $\sigma(u, u) > 0$ for any $u \neq 0$.

This is similar to the Hermitian form which arose in the context of Neumann stability and the discrete Fourier interpolation in Section 4.3. At that time the form was defined on the complex vector space $\mathbb{C}^n$ via certain function values. In another direction, $\sigma(u, u)^{1/2}$ defines a norm [Loustau and Dillon

(1993)] as introduced in Section 2.2.

The particular inner product of interest is $\sigma(f,g) = \int_a^b fgdx$, defined on the space of continuous functions of an interval. See also Exercise 4.

Returning to the current question, the interpolating polynomials are called the Legendre polynomials. This set of polynomials forms a basis for the space $\mathbb{P}_n$ of the polynomials of degree no larger than n. Recall that the degree n Lagrange polynomials also form a basis of $\mathbb{P}_n$. In addition, the Lagrange polynomials satisfy the relation $l_i(x_j) = \delta_{i,j}$ for interpolation points x_i. The determining characteristic for the Legendre polynomials is $\sigma(p_i, p_j) = \int_a^b p_i p_j = \delta_{i,j}$. Hence, the Legendre polynomials are orthogonal for this inner product σ. More generally, we will refer to any pair of integrable functions on $[a, b]$ as orthogonal provided $\int_a^b fg = 0$. For instance, $f(x) = x$ and $g(x) = 1$ are orthogonal for the interval $[-1, 1]$.

We start with a function f defined on an interval $[a, b]$ with a partition $a \leq x_0 < x_1 < ... < x_n \leq b$, and consider Hermite interpolation of f.

$$h(x) = \sum_{i=0}^{n} f(x_i)H_i(x) + \sum_{i=0}^{n} f'(x_i)S_i(x), \tag{5.3.1}$$

where each H_i and S_i is a polynomial of degree $2n+1$ and $H_i(x_j) = \delta_{i,j}$, $S_i(x_j) = 0$ while $dH_i/dx(x_j) = 0$, $dS_i/dx(x_j) = \delta_{i,j}$. In section 3.5, we proved that the Hermite interpolation exists for any n. Further, we have derived an error estimate for this interpolation. In particular, for $p_{n+1}(x) = \prod_{i=0}^{n}(x - x_i)$,

$$f(x) - h(x) = e(x) = \frac{f^{(2n+2)}(\xi_x)}{(2n+2)!} p_{n+1}(x)^2, \tag{5.3.2}$$

provided f is $2n+1$ times continuously differentiable.

The next step is to use the Hermite interpolation to estimate the integral of f.

Definition 5.3.1. Let f be a $C^{2n+2}[a, b]$ function. Then the *Hermitian quadrature* for the integral of f relative to a partition $a \leq x_0 < x_1 < ... < x_n \leq b$ is

$$\sum_{i=0}^{n} \gamma_i f(x_i) + \sum_{i=0}^{n} \delta_i f'(x_i), \tag{5.3.3}$$

where the coefficients are given by

$$\gamma_i = \int_a^b H_i = \int_a^b [1 - 2h_i'(x_i)(x - x_i)]h_i, \;\; \delta_i = \int_a^b S_i = \int_a^b (x - x_i)h_i,$$

with $h_i = p_i^2$. The nonzero coefficients are called the *weights*.

As a consequence of (5.3.1) and (5.3.2), we estimate the *Hermitian quadrature error.*

Theorem 5.3.1. *The error E for the Hermite quadrature is bounded by*

$$|E| \leq \frac{M}{(2n+2)!} \int_a^b p_{n+1}^2 \leq \frac{M}{(2n+2)!}(b-a)^{2n+3}, \tag{5.3.4}$$

where M denotes $\max_{x\in[a,b]} |f^{(2n+2)}|$. Furthermore, if f is a polynomial of degree less than or equal to $2n+1$, then the estimate for the integral of f given by (5.1.3) is exact.

Proof. It is only necessary to integrate the error estimate for Hermite interpolation and then notice that $|p_{n+1}^2(x)| \leq (b-a)^{2n+2}$. Finally, the last assertion follows from the statement, $f^{(2n+1)} = 0$ provided $f \in \mathbb{P}_{2n+1}$. □

Next, we want to look at Hermitian quadrature with the additional assumption that $\delta_i = 0$ for each i. We see that this requirement corresponds to a particular choice of the x_i. For this purpose, we use the Lagrange polynomials. Recall that for each $i = 0, 1, ..., n$, the ith Lagrange polynomial is given by

$$l_i(x) = \frac{\prod_{j\neq i}(x - x_j)}{\prod_{j\neq i}(x_i - x_j)}.$$

These polynomials form a basis for $\mathbb{P}_n$. In addition, $l_i(x) = p_{n+1}(x)/[p'_{n+1}(x_i)(x - x_i)]$.

We calculate,

$$0 = \delta_i = \int_a^b (x - x_i)h_i = \int_a^b (x - x_i)l_i^2 \tag{5.3.5}$$

$$= \int_a^b (x - x_i)\left(\frac{1}{x - x_i}\frac{p_{n+1}(x)}{p'_{n+1}(x_i)}\right)^2 = \frac{1}{p'_{n+1}(x_i)}\int_a^b p_{n+1}(x)l_i(x),$$

since $l_i = p_{n+1}(x)/[p'_{n+1}(x_i)(x - x_i)]$. Hence, the condition $\delta_i = 0$ implies that p_{n+1} is orthogonal to the space of all polynomials of degree no larger than n. We state this formally in the following theorem.

Theorem 5.3.2. *Suppose that $f \in C^{2n+2}$, and p_{n+1} is orthogonal to $\mathbb{P}_n$, the polynomials of degree no larger than n, then the Hermitian quadrature for f is given by the weighted sum of function values,*

$$\int_a^b f = \sum_{i=0}^n \gamma_i f(x_i) + E; \;\; \gamma_i = \int_a^b H_i; \;\; |E| \leq \frac{M}{(2n+2)!}\int_a^b p_{n+1}^2. \tag{5.3.6}$$

Proof. This result is a simple consequence of Theorem 5.3.1 and (5.3.3) and (5.3.5). Note that (5.3.5) actually states that p_{n+1} is orthogonal to a basis for $\mathbb{P}_n$. It is immediate that any vector orthogonal to a basis of a space is orthogonal to the entire space. □

The special case that arose in Theorem 5.3.2 is called Gaussian quadrature.

Definition 5.3.2. The numerical integration procedure that arises when $p_0, p_1, ..., p_{n+1}$ are orthogonal is called *Gaussian quadrature.*

The next step is to determine when p_{n+1} is orthogonal to the space $\mathbb{P}_n$. As the Lagrange polynomials form a basis for this space, then we need only consider the integrals $\int l_i p_{n+1} = 0$. Given $0 = \int_a^b S_i = \int_a^b (x - x_i) l_i$ and setting $f = l_j$ in (5.3.6), we compute

$$\int_a^b l_j = \sum_{i=0}^{n} l_j(x_i) \int_a^b H_i = \int_a^b H_j = \int_a^b u_j h_j$$

$$= \int_a^b [1 - \frac{d}{dx} h_j(x_j)(x - x_j)] l_j^2 = \int_a^b l_j^2 - \frac{d}{dx} h_j(x_j) \int_a^b S_j = \int_a^b l_j^2.$$

We have now proved the following corollary.

Corollary 5.3.1. *The polynomial* $p(x) = \prod_{i=0}^{n} (x - x_i)$ *is orthogonal to* $\mathbb{P}_n$ *provided the Lagrange polynomials satisfy*

$$\int_a^b l_i = \int_a^b l_i^2. \tag{5.3.7}$$

In this case the weights are given by $\int_a^b l_i > 0$.

Hence, *Gaussian quadrature* is the special case of Hermite quadrature associated to (5.3.6), and the polynomials p_{n+1} are the *Legendre polynomials.* Gaussian quadrature with n points has the property that it is exact for polynomials of degree $2n + 1$. There is no numerical integration technique that is better. It all depends on the choice of the points $x_0, x_1, ...,$ the partition points. This choice for these points is commonly called *Gaussian quadrature points*. In turn, to find the Gaussian quadrature points for a given n we must identify the Legendre polynomial p_n as these points are the roots.

To identify the Gaussian quadrature points, we first determine the sequence of Legendre polynomials, the orthogonal polynomials. The points are the roots of the polynomials. We start with $p_0 = 1$ on $[a, b]$ and then use the Gram-Schmidt process to solve for each p_n in succession. For instance, for the interval $[-1, 1]$, $p_1(x) = x$ and $p_2(x) = x^2 - x + 1/6$. The corresponding Gaussian quadrature points for this interval are 0 for $n + 1 = 1$ and $0.5 \pm 0.5/\sqrt{3}$ for $n + 1 = 2$.

Additionally the points and weights for the standard interval $[-1, 1]$ are listed on line or in pre-computer era numerical analysis texts [Hildebrand (1974)]. On the other hand, the *Mathematica* function *GaussianQuadrature*$[n, a, b]$ returns the Gaussian quadrature points and weights for the nth order quadrature on the interval $[a, b]$. In the literature, it is possible to find approximations for the Gaussian point for n in the tens of thousands.

Exercises:

1. Given a set of Gauss points for an interval $[a, b]$ determine the transformation that will map them to the corresponding Gaussian quadrature points another interval $[\alpha, \beta]$.

2. Consider two point Gaussian quadrature on the interval $[0, 4]$ and the function $f_n = x^n + 2x - 3$. For $n = 4, 5, 6$,
a. compute the Gaussian quadrature estimate for $\int_0^4 f dx$,
b. compute the actual error and the estimated error.

3. Let $f(x) = xe^{-x} - 1$ and set the interval to $[1, 4]$. Compute the integral of f using the two point Gaussian quadrature. Compare these results with those of Exercises 1 and 2 of Section 5.1 and Exercise 1 of Section 5.2.

4. Consider the vector space $C^0[a, b]$ of continuous real valued functions on the interval $[a, b]$. Prove that the function $\sigma(f, g) = \int_a^b fg dx$ is a positive definite inner product on the vector space $C^0[a, b]$.

5. For the interval $[-1, 1]$, set $p_0(x) = 1$ and use the Gram-Schmidt process to derive p_1 and p_2, so that p_0, p_1 and p_2 are orthogonal.

6. Determine the Legendre polynomial p_1 or an arbitrary interval. Prove that one point Gaussian quadrature is identical to the midpoint rule.

7. Prove that there are Gaussian quadrature points for every n. This is not the case for some of the weighted quadratures introduced in the next section.

5.4 Comments on Numerical Integration

In this section, we resolve three unrelated numerical integration questions.

We begin with weighted quadratures. This is a generalization of Gaussian quadrature which addresses certain functions that do not conform well to general techniques. Even though Gaussian quadrature is the best of the numerical integration procedures, there are functions for which the procedure does not perform well. In particular, that Gaussian quadrature performs well for polynomials, but what of functions with properties that are decidedly not *polynomial-like*? The weighted quadratures are intended to address these cases. We begin the section with the question.

In the second, we look at functions that are represented in parametric form. We address the problem of applying numerical integration to these functions. The problem here is that we must rewrite a parametric representation, $\gamma(t) = (\gamma_1(t), \gamma_2(t))$ into the format $(x, f(x))$.

Lastly we look at applying Gaussian quadrature to rectangles and triangles. For the rectangle case, we can use Cartesian products. For triangles, there is a transformation between rectangles and triangles that enables gaussian quadrature via a change of variables.

There are many different realizations of quadrature. Gaussian quadrature is perhaps the simplest form. The idea is that Gaussian quadrature is exact for polynomials up to a degree associated to n, the number of points. In particular, it is well-behaved for functions with bounded derivatives $\{f^{(m)}(x),\ m = 1, 2, ..\}$. Hence, the error term goes to zero as n increases. Specifically, the more points used in the numerical process, the better the approximation. Conversely, it can be that f is continuous on a closed interval but the derivatives of f do not form a bounded set. There are other cases where the function is asymptotic, not a property that is shared by polynomials. The weighted quadratures are intended to resolve these cases.

We begin this section with *Chebyshev quadrature*. In this case we are concerned with a function f defined on an interval $[-1, 1]$ so that $f^{(n)}(x)$ is not bounded as $x \to -1$ or $x \to 1$ or both. Stated otherwise, the function has vertical tangent at one or the other end point. It is reasonable to look at $g(x) = f(x)/\sqrt{1 - x^2}$. We refer to g as the weighted form of f. We expect that the denominator resolves the anomaly. Hence, we can apply a quadrature procedure to g. With the notation of the last section, we follow the direction presented in (5.3.3) and (5.3.6) modified now for the weighted case. In particular, we want to know when the $\delta_i = 0$ and in this case what we can say about E and γ_i. To begin, we state a lemma. Note that the inner product for this case is the weighted product,

$$\sigma(f,g) = \int_a^b \frac{fg}{\sqrt{1-x^2}}.$$

Lemma 5.4.1. *With the current notation, $\delta_i = 0$ if and only if p_{n+1} satisfies*

$$0 = \int_{-1}^{1} \frac{p_{n+1}q}{\sqrt{1-x^2}} dx, \tag{5.4.1}$$

for any $q \in \mathbb{P}_n$.

Proof. The proof is entirely analogous to the corresponding result for Gaussian quadrature, Theorem 5.3.2. □

Next, we resolve (5.4.1) with a variable substitution $x \to \cos(\theta)$. It is then routine to verify the following statements.

Theorem 5.4.1. *The Chebyshev quadrature satisfies*

$$p_{n+1}(x) = C_{n+1}\cos((n+1)\cos^{-1}(x)),$$

$$E = \frac{\pi}{2^{2n+2}(2n+2)!} f^{(2n+2)}(\xi), \tag{5.4.2}$$

where $|\xi| < 1$ and C_{n+1} is a constant. The weights satisfy $\gamma_i = \pi/(n+1)$.

If we compute the points x_i, we find that they are clustered at the outer edges of the interval. For instance, see Exercise 3.

Next, we turn to *Laguerre quadrature.* In this case we consider the half line $[0,\infty)$ and the weight e^{-x}.

Theorem 5.4.2. *The Laguerre quadrature satisfies*

$$p_n(x) = \frac{e^x}{n!}\frac{d^n}{dx^n}(e^x x^n); \quad E = \frac{((n+1)!)^2}{(2n+2)!} f^{(2n+2)}(\xi), \tag{5.4.3}$$

where $0 < \xi < \infty$. The weights satisfy $\gamma_i = (i!)^2$. The p_n are called the Laguerre polynomials.

Proof. See Exercise 5 □

Another important case is the Jacobi quadrature. This is a generalization of the Chebyshev case. In this case the weighting function is $w(x) = (1-x)^\alpha(1+x)^\beta$, $\alpha, \beta > -1$ Jacobi quadrature is useful for numerical solutions to PDE with fractional derivatives. These PDE arise for anomalous diffusion.

In previous sections, we have developed procedures to approximate an unknown function by parametric cubics. For these cases, the implementation of the trapezoid method or any other of the numerical integration techniques requires preprocessing.

Consider the following setting. Suppose we have a function $f(x)$ that is known only at a few domain values. Suppose further that we use these values to fit B-spline segments $\gamma^k(t) = (\gamma_1^k(t), \gamma_2^k(t))$, $t \in [0,1]$ associated to points on the graph of f. Notationally we designate the points $(x_i, f(x_i))$, $i = 0, 1, ..., n$. Next, given points in $a \le x_0 \le ... \le x_n \le b$ in the domain of f, we want to find y_i on the B-spline so that

$$\int_a^b f dx \approx \sum_{i=1}^{n} \frac{y_i + y_{i-1}}{2}(x_i - x_{i-1})$$

approximates the integral of f or equivalently the area under the B-spline?

We can rephrase this equation as follows. Given a point (x, y) on the B-spline, how can we find k and t so that $(x,y) = (\gamma_1^k(t), \gamma_2^k(t))$? If we can resolve this, then we can write (x_i, y_i) as $(\gamma_1^{k_i}(t_i), \gamma_2^{k_i}(t_i))$ and approximate the integral with

$$\sum_{i=1}^{n} \frac{\gamma_2^{k_i}(t_i) - \gamma_1^{k_i}(t_{i-1})}{2}(\gamma_1^{k_i}(t_i) - \gamma_1^{k_j}(t_{i-1})), \tag{5.4.4}$$

$j = i$ or $i-1$. Hence, we must solve for k, the segment identifier, and t, the parameter value.

To determine k, we must find k such that $\gamma^k(0) \le x \le \gamma^k(1)$.

To locate t, we use Newton's method via the FindRoot function.

```
FindRoot[gamma^k[t] == x_i, {t, tau_i}];
```

where τ_i is the estimate for t_i required by the method. We claim that $\tau_i = t_{i-1}$ is a reasonable estimate provided x_i and x_{i-1} lie in the same B-spline segment. Alternatively, we take $\tau_i = 0$. We base these choices on the tacit assumption that the differences $x_i - x_{i-1}$ are small.

Keep in mind that Newton's method must always be suspect, as it may produce a t_i, which is not between 0 and 1. Therefore, it is best to test the value returned for reasonableness. In particular, we expect t_i to be in $[0, 1]$, and we expect that t increases as x increases. The following pseudo code implements these requirements.

```
t_i = FindRoot[ x[t] == x_i, {t, \tau_i]}];
If [t_{i-1} > t_i || t_i > t_n,
    exit and print an error statement,
    otherwise continue with processing
];
```

We consider a third question. Suppose we want to integrate a function over a triangle. This is important for numerical differential equations, finite element method. In that context, we may need to compute millions of integrals over triangles for a single problem. We will need to do that efficiently and accurately. The problem is that integrals over triangular regions are not easy. Even *Mathematica* does not do this well.

The Gaussian quadrature developed in the prior section applies only to intervals. But it is routine to extend to rectangles via Cartesian products. For instance two point quadrature on an interval becomes 4 point quadrature on a rectangle. Given Gaussian quadrature points $\{x_1, x_2\}$ on the x-axis interval $[a, b]$ and $\{y_1, y_2\}$ on the y-axis interval $[c, d]$, we merely take the Cartesian product $\{x_1, x_2\} \times \{y_1, y_2\}$ to get Gaussian quadrature points for the rectangle, $[a, b] \times [c, d]$. The weights for the x direction interval are $(b - a)/2$ at both locations. In the y direction, we have $(d - c)/2$. The corresponding weights for the rectangle are $(b - a)(d - c)/4$. The idea is to apply Fubini's theorem,

$$\int_a^b \int_c^d f(x, y)dydx = \int_a^b [\int_c^d f(x, y)dy]dx$$

$$\approx \sum_{j=1}^{2} \frac{d-c}{2} \int_a^b f(x, y_j)dx \approx \sum_{i=1}^{2} \sum_{j=1}^{2} \frac{(b-a)(d-c)}{4} f(x_i, y_j). \qquad (5.4.5)$$

Now, suppose you want to integrate over a triangular region E with vertices $(\alpha, \beta), (\gamma, \delta), (\mu, \nu)$. The rational function in two variables

$$\Phi(x, y) = (\xi, \eta) = \left(\frac{(1+x)(1-y)}{4}, \frac{1+y}{2}\right)$$

maps the rectangle $[-1, 1] \times [-1, 1]$ to the triangle $(0, 0), (1, 0), (0, 1)$. Furthermore, it is bijective on the interior of the rectangle. It is not bijective on the boundary, but that is not important for the integral. Continuing, the affine mapping $A(x,y) = T(x,y) + (\alpha,\beta)$, where T is the linear transformation given by the matrix

$$\begin{pmatrix} \gamma - \alpha & \mu - \alpha \\ \delta - \beta & \nu - \beta \end{pmatrix},$$

will map the triangle $(0,0),(1,0),(0,1)$ bijectively to the triangle E, with vertices (α,β), (γ,δ), (μ,ν). Hence, by combining A and Φ, we have a bijection from the standard rectangle to any triangle. Therefore, employing the usual variable change procedure [Marsden and Tromba (2003)], we are able to rewrite

$$\int_E f(\xi,\eta)dA = \int_{-1}^{1}\int_{-1}^{1} f \circ A \circ \Phi(x,y) J(A \circ \Phi) dx dy.$$

For the final step, approximate the rectangular integral via the quadrature technique (5.4.5).

Exercises:

1. Prove Lemma 5.4.1.

2. Complete the verification of the statements for Theorem 5.4.1.

3. For Chebyshev quadrature compute the roots of p_{21}, quadrature points x_i $(n = 20)$.

4. The 5th Laguerre polynomial is $120-600x+600x^2-200x^3+25x^4-x^5$. Approximate the roots of L_5. Describe how they are distributed.

5. Prove Theorem 5.4.2.

a. Derive the expression for the Laguerre polynomials from

$$0 = \int_0^{\infty} e^{-x} p_{n+1} q dx,$$

for $q \in \mathbb{P}_n$.

b. Derive the expression given E.

c. Find H_n and then derive the expression for γ_i.

6. Let $f(x) = xe^{-x} - 1$ and consider the guide points $(1, f(1))$, $(1.5, f(1.5))$, $(2, f(2))$, $(2.5, f(2.5))$, $(3, f(3))$, $(3.5, f(3.5))$ and $(4, f(4))$.

a. Use a B-spline to fit this set of guide points. Plot the resulting curve.

b. Beginning with the B-spline, use the trapezoid method to approximate the integral of f. Compare this result with the output of Exercises

1 and 2 of Section 5.1, Exercise 1 of Section 5.2 and Exercise 3 of Section 5.3.

7. Compute the arc length of the B-spline derived in Problem 6(a) above. In this case the curve is the join of four B-spline segments, σ^i, $i = 1, 2, 3, 4$. Compute the length of each B-spline segment individually using a uniform partition with $\Delta t = 0.1$.

8. Use the quadrature technique of this section to estimate

$$\int_E x^4 + 2xy + y^4 dxdy$$

where E is the triangle with vertices (-1, 2), (5, -1), (0, 4).

9. Give an example of a function for which $f^m(\xi)/m!$ does not converge to zero as m gets large.

10. Repeat Exercise 2 of Section 2. Except this time we will use cubic B-splines to interpolate the given values.

a. Plot the 26 points $(t, \delta(t))$. Fit a cubic B-spline to the data and display both plots on the same axis. In order to ensure that the curve extends near to the domain end points, duplicate the first and last points when generating the B-spline. We denote this curve σ, and the segments of σ as $\sigma^i(s) = (\sigma_1^i(s), \sigma_2^i(s))$.

b. The temperature difference δ is caused by the reaction. The first t for which $\delta(t) > 0$ is the starting time of the reaction, denoted a. To find the end time of the reaction plot the points $(t, \log(\delta(t))$. You will notice that after a while the points seem to lie on a line. The value of t for which the plot begins to appear linear is denoted b and is the end time for the reaction.

c. Estimate the integral of δ on $[a, b]$. Use the midpoint rule for the partition determined by subdividing the interval into 10 sub-intervals with length $h = (b - a)/10$.

d. Estimate the integral $\int_a^b \delta(t)dt$ using the trapezoid method. Use the same partition as in Part c.

e. Recall from calculus that $|\int_a^b \delta(t)dt| \leq \max_{t \in [a,b]} |\delta(t)|(b - a)$. Use the B-spline to estimate the maximum of δ.

11. Consider the integral $\int_E x^2 + y^2 dA$ where E is the triangle with vertices $(1, 1)$, $(2, 4)$, $(4, 4)$. Compute the integral directly as a double

integral and compute the integral using the two point Gaussian quadrature for the rectangle $[-1, 1] \times [-1, 1]$.

Chapter 6

Numerical Ordinary Differential Equations

Introduction

In this chapter our primary concern is first order ordinary differential equation, ODE, initial value problems. In particular, a first order ODE is an expression

$$\frac{du}{dx} = u' = f(x, u(x)), \quad x \geq x_0 \tag{6.0.1}$$

with an initial value,

$$u_0 = u(x_0). \tag{6.0.2}$$

The basic idea is that we know the derivative f, the initial location or time, x_0, and the initial function value $u(x_0)$, Hence, we know $f(x_0, u(x_0)) = u'(x_0)$. Now, with this information and Taylor's theorem, we can begin to approximate incremental values for u. In this chapter, we identify three procedures to estimate values $u(x_n)$ for a sequence $x_0, x_1, \ldots$.

There is an existence, uniqueness theorem for first order ODE. This theorem requires the Lipschitz condition. Recall that the Lipschitz condition arouse in Chapter 3 in conjunction with B-splines. This result is due to Picard. It is classical.

Theorem 6.0.1. *Consider the first order ODE, $u' = f(x, u)$ with given initial value $u_0 = u(x_0)$ and where f is a continuous function of its domain D and (x_0, u_0) is interior to D. Suppose that f satisfies the Lipschitz condition on D. In particular, suppose that there is a non-negative real K so that*

$$|f(x, y_2) - f(x, y_1)| < K|y_2 - y_1|,$$

for any pair of points (x, y_1) and (x, y_2) in D. Then there is an interval I about x_0 so that the ODE has unique solution on I.

Proof. See [Simmons and Robertson (1991)] for a proof. □

A word of caution is necessary. The techniques developed in this chapter apply to any ODE. It is not difficult to write down an equation for which there is no solution or alternatively an equation without unique solution. In either case, the results of the numerical method may not represent the setting which has been modeled. It is important to stay within the context of the theorem.

The techniques we develop are similar to the FDM time stepping procedures we introduced in Chapter 4. Indeed, if the independent variable is time, then the iterated steps are time stepping. This is the source of the expression, *initial value problem.* In the last section, we tie ODE methods to FDM.

In Section 3, we look at second order ODE. We will see that second or higher order equations can be reduced to a system of first order equations. For instance, consider a second order equation with two initial values,

$$u'' = g(x, u', u); \;\; u_0 = u(x_0), \;\; u_0' = u'(x_0).$$

It is easy to see that this equation may be recast as a pair of first order initial value problems. In particular, we set $v = u'$ and write the pair of equations

$$v' = g(x, v, u), v_0 = v(x_0); \quad u' = v, u_0 = u(x_0).$$

We see that the techniques we developed here for order 1 equations are routinely extended to sets of order 1 equations and thereby apply to equations of order one or two.

In another direction, suppose we have a second order ODE where u is defined on an interval $[a, b]$. If we specify values $u(a)$ and $u(b)$ then we say that we have a *boundary value problem.* We will approach these problems with a technique called the *shooting method.* The basic idea is, given $u(a)$, determine $u'(a)$ so that u has the desired value at b.

In addition to the pure ODE methods introduced here, there are hybrid techniques. We will look at the *method of lines.* In this case, we develop the spatial part of a PDE via FDM and then apply ODE techniques to the time stepping process. Hybrid techniques are very common in the literature. The method of lines in only one instance.

We begin the chapter with three initial value ODE techniques, forward Euler, forward Euler as predictor/corrector and midpoint method. The predictor/corrector uses a given technique is a predictor. The corrector is a related process that modifies the value returned by the predictor. If the two values are sufficiently close, then we proceed to the next step using the current predictor value. Otherwise, we have the option to repeat the corrector and compare the two corrected values. When successive values of the corrector are sufficient close, then we proceed with the current corrector value. For this procedure, there is a tacit assumption that the sequence of correctors converges to the actual derivative of u at the next x. At the end of the chapter we connect this question to one about the Crank Nicolson method FDM.

If we continued beyond forward Euler and midpoint, we would come to Runge-Kutta method. This method is certainly important. However, the mathematical foundation is too detailed for this treatment. See [Loustau (2016)] for Runge-Kutta or R-K method.

Following the initial value methods, we develop the shooting method for second order ODE. This is an example of a boundary value method. We end the chapter with a treatment of the method lines. This is the hybrid technique that was referred to above.

A matter of note is that the case studies of this chapter are mostly population studies. We will see two birth death models. In the second case we look at population model that arises in *mathematical oncology.*

6.1 First order ODE Techniques, Forward Euler and Corrector Method

We begin with a first order ODE, $u' = f(x, u)$ and an initial value $u(x_0) = u_0$. Given Δx and $x_n = x_0 + n\Delta x$, we seek a means to approximate $u_n = u(x_n)$. Our first technique is called *forward Euler.* In this case, we set $u_1 = u_0 + u'(x_0)\Delta x = u_0 + f(x_0, u_0)\Delta x$. More generally, we have an iterating process,

$$u_{n+1} = u_n + f(x_n, u_n)\Delta x. \tag{6.1.1}$$

We have encountered the terminology *forward Euler* in another context. This instance is related. We return to this in Section 4.

Consider an example. We begin with an ODE, one, we know describes an exponential function. In particular, we take $u' = f(x, u(x)) = \alpha u(x)$ with $x_0 = 0$, and $u_0 = u(0) = 1$. In addition, we choose $\Delta x = 0.1$. Hence, for $\alpha = 2$,

$$u_1 = u_0 + f(x_0, u(x_0))\Delta x = u_0 + \alpha u_0 \Delta x = 1 + 2 * 1 * 0.1 = 1.2.$$

The actual solution to the equation is $u(x) = e^{\alpha x}$. Therefore, $u(x_1) = e^{2*0.1} \approx 1.221$ and the error is about 0.021.

In turn, we calculate

$$u_2 = u_1 + f(x_1, u(x_1))\Delta x = u_1 + \alpha u_1 \Delta x = 1.2 + 0.24 = 1.44,$$

whereas $u(x_2) = e^{2*0.2} \approx 1.492$. Now the error is about 0.052, more than twice the error after the first step. Continuing in this manner, each successive approximation is worse. In fact, the error appears to grow without bound. (See Exercise 1.) In Chapter 4, we would refer to this as unstable.

There is an alternative approach. We are using $f(x_0, u_0)$, the slope at x_0 to approximate the slop at x_1. However, we can take advantage of the approximate value u_1 and compute a second estimate of the slope for u at x_1 using $f(x_1, u_1)$. In particular, we compute the average $1/2(f(x_0, u_0) + f(x_1, u_1))$ and compute an alternate value for u using this estimate for the slope. We denote this value $u_{11} = u_0 + 1/2(f(x_0, u_0) + f(x_1, u_1))\Delta x$ is the *corrector* for u_1.

For instance, suppose we continue with the current example. In this case $u_1 = 1.2$ and $((f(x_0, u_0) + f(x_1, u_1))/2 = (2 + 2.4)/2 = 2.2$. Proceeding,

$$u_{11} = u_0 + 2.2 * \Delta x = 1 + 0.22 = 1.22.$$

This is correct to two decimal places.

Note that if $|u_1 - u_{11}| \gg 0$, then we will want to apply the corrector a second time to calculate u_{12}. Again we compute $|u_{11} - u_{12}|$ and decide if a third corrector is needed. In the end, we use u_{1n} to approximate u at x_1.

Which value should we use to compute x_2? Since we report u_{1n} as the estimated value of u at x_1, then we should use it to compute u_2.

Generally speaking, the forward Euler is not considered to be a good approximation procedure. The status is much the same as with forward Euler FDM. The forward Euler with corrector is better but not efficient. And the more often you execute the corrector, the less efficient it becomes. For instance, if you execute the corrector logic multiple times at each step, then the execution time increases by a fixed multiple while the magnitude of the data improvement decreases with each corrector iteration. Hence, multiple executions of the corrector is not often advised.

We end the section with another example. Suppose $N = N(t)$ represents a population size at time t. Consider the following population problem. The net change in N relative to t should be the number births minus the number of deaths plus or minus migration effects. The births may be estimated at the birth rate times the population size. Whereas, the death rate will be dependent on environmental factors (for instance, temperature, available of food, infectious agents in the environment). Suppose that there is an experiment during which migration is prevented, the birth rate is known to be 0.09 and the death rate as a function of time and population size, is given by $B(t)N^{1.7}$. In particular,

$$N'(t) = 0.09 * N(t) - B(t)N(t)^{1.7} = f(t, N(t)). \tag{6.1.2}$$

In addition, suppose that $B(t)$ is known for 11 equally spaced incremental values of t. These values are given in the following table.

t	0	1	2	3	4
B	0.007	0.0036	0.0011	0.0001	0.0004

5	6	7	8	9	10
0.0013	0.0028	0.0043	0.00056	0.00044	0.0004

This type of population model is called a *birth/death model.* It is the simplest of the population models.

Finally, we are given $N(0) = 100$, the time interval $[0, 1]$ and $\Delta t = 0.1$. Hence, the values in the first row of the table now refer to integer multiples of Δt. We must use forward Euler and the corrector to estimate the population at each of the given times.

Looking at the equation (6.1.2), we observe that population growth is exponential while decline is polynomial. Even though the rate of growth is small, the values for B are also small. Hence, we should expect to see overall growth during the duration. We first estimate the population over the time period using the basic Euler method. Since we expect integer values of N, we round the computed values at each step. We compute

$$f(0, 100) \approx -8.49, N(0.1) = 99; \;\; f(0.1, 99) \approx 0.02, N(0.2) = 99,$$

$$f(0.2, 99) \approx 6.19, N(0.3) = 100; \;\; f(0.3, 99) \approx 8.75, N(0.4) = 101.$$

We continue through 11 time values and report output in Figure 6.1.1. As we expect integer output, we round the computed data at each time step.

Next we employ the corrector. The first few calculations are summarized here. We use g for the derivative estimate by means of the corrector.

$$g(0, 100) \approx -8, 58, N(0.1) = 99; \;\; g(0.1, 99) \approx 0.02, N(0.2) = 99,$$

$$g(0.2, 99) \approx 6.21, N(0.3) = 100; \;\; g(0.3, 99) \approx 8.79, N(0.4) = 101.$$

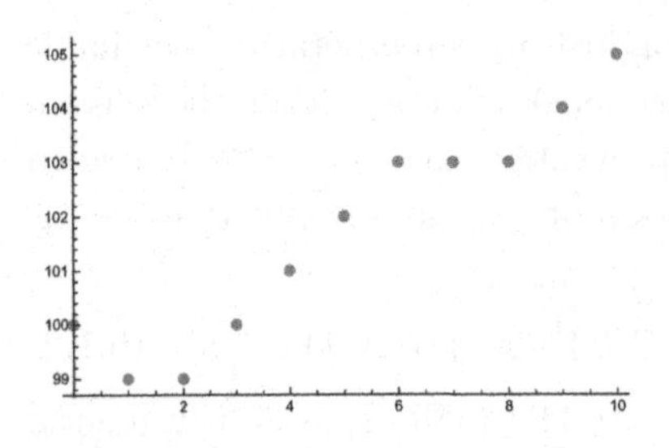

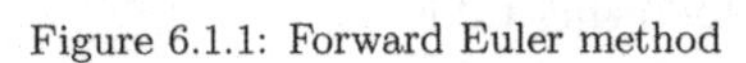

Figure 6.1.1: Forward Euler method

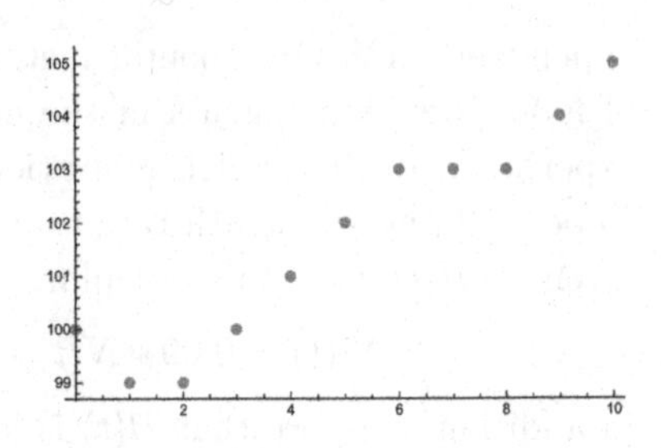

Figure 6.1.2: Forward Euler with corrector

The estimated derivatives are slightly changed, but the population figures shown in Figure 6.12 are also rounded to the nearest integer and unchanged.

Exercises:

1. Continue the example developed above, $u' = \alpha u(x)$ with $x_0 = 0$, $u_0 = 1$ and $\Delta x = 0.1$. Given that $u(x) = e^{2x}$, use forward Euler to compute the estimates $u_3, ..., u_{10}$. For each case, compute the corresponding error.

2. Apply forward Euler to the equation given by (6.1.2).

3. Repeat 2 for Euler with a single corrector.

4. Repeat 2 for Euler with 3 correctors. How does this data change with each iteration?

5. Repeat Exercise 2 with the following changes.
a. Use least squares to fit a cubic polynomial to the given values for B. (See Figure 6.1.3.)
b. Using the cubic to represent B and $\Delta x = 0.05$, estimate values of N. Note that the additional values for B provided by the B-spline are essential here.

6.2 Midpoint Method with an Application to Mathematical Oncology

In Section 6.1, we considered the following equation, $u' = \alpha u$ and an initial value $u(0) = 1$. This equation has actual solution $u = e^{\alpha x}$. In addition, we took $\Delta x = 0.1$. At that time, we used the forward Euler method and computed approximate values for u. Next, we considered a corrector method and observed that the corrected values were much better than the initial approximations. Alternatively, we saw another example where the results were the same whether or not the corrector was used.

In this section, we develop the midpoint method. This is the third ODE technique. As a general rule, midpoint is more efficient than the predictor/corrector while producing higher quality results than the forward Euler alone.

We unify the notation for the three techniques as follows. Given an ODE, $u'(x) = f(x, u(x))$, we write the iterated procedure $u_{n+1} = u_n + h(x, u(x))\Delta x$. For each method, we resolve h differently. For forward Euler, $f = h$. For the corrector, $h(x_n, u_n) = 1/2(f(x_n, u_n) + f(x_{n+1}, u_{n+1}))$. For midpoint method, the procedure is described now.

Recall Equation (6.1.1) where we use $f(x_n, u_n)$ to approximate the derivative of u at x_{n+1}. For the midpoint method, we compute u_{n+1} from $x_n + \Delta x/2$ and $u_n + f(x_n, u_n)\Delta x/2$. In particular, the *midpoint method* is given by

$$u_{n+1} = u_n + f\big(x_n + \frac{\Delta x}{2}, u_n + f(x_n, u_n)\frac{\Delta x}{2}\big)\Delta x. \tag{6.2.1}$$

Note the similarity to Crank Nicolson FDM. In that case, you create a fictitious time step $t_{n+1/2}$. In this case, we evaluate a value of f at a point midway between x_n and x_{n+1}. In a sense, we are creating a fictitious value of the dependent variable, $u_{n+1/2}$. We do this using the Taylor expansion for u at $x_{n+1/2}$. We see in Section 4 that this is not actually related to Crank Nicolson.

Returning to the ODE $u' = 2u$ with $u(0) = 1$ and $\Delta x = 0.1$ and using (6.2.1)

$$u_1 = u_0 + f(x_0 + \Delta x/2, u_0 + \alpha * u_0 * \Delta x/2)\Delta x$$

$$= 1 + \alpha(1 + \alpha * 0.05) * 0.1 = 1.22,$$

and

$$u_2 = u_1 + f(x_1 + \Delta x/2, u_1 + \alpha * u_1 * \Delta x/2)\Delta x$$

$$= 1.22 + 2(1.22 + 2 * 1.22 * 0.05) * 0.1 = 1.4884.$$

Recall that $u(1) = 1.2214$ and $u(2) = 1.49182$. We see that midpoint values are very good.

Runge Kutta method is significantly better. This is the next stage in ODE simulation technique. Not only is it more correct, Runge Kutta includes an error estimation procedure. [Loustau (2016)] However, this method is best presented at a later stage.

Scientists use birth/death processes to model tumor growth. In this case, cell division is governed by the mitosis rate. This is exponential. But tumors rarely grow exponentially. Rather, only the cells on the tumor surface have access to sufficient nutrients via the blood supply to divide. Hence, the growth rate is a function of the tumor surface area or in the case of a more or less spherical tumor, the tumor radius. Furthermore, there is cell death. Hence, we should expect a birth/death model. As with the population of dividing cells, dying cells can be identified geometrically. Indeed, cell death should be more prevalent with older cells, those lying closer to the tumor center. Hence, both processes are functions of a fraction of cell population. Researchers often consider the following model. [Wodarz and Komarova (2014)]

$$\frac{dU}{dx} = aU^{\alpha} - bU^{\beta}, \tag{6.2.2}$$

where specific values for the parameters a, b and α, β are to be determined through experimental observation. These parameters identify subpopulations and rates of change. In particular, U^{α} is the subpopulation of dividing cells and a is the mitosis rate, U^{β} is the aging subpopulation and b is the death rate.

There are parameter values for which U is known. For instance, $U_t = aU - bU^2$ has solution $U = KU_0e^{at}/(K + U_0(e^{at} - 1))$ where $K = a/b$ is called the *carrying capacity*. Using l'Hopital, $\lim_{t\to\infty} U = K$. Hence, U is increasing taking values on the interval $[U_0, K)$ for $t \in [0, \infty)$. The resulting curve is called *sigmomial.* We have plotted U in Figure 6.2.1 for the case $K = 2$.

For the ODE given in (6.2.2), f is only a function of U and not directly a function of t. Hence, (6.2.1) becomes

$$U_{n+1} = U_n + f\big(U_n + f(U_n)\frac{\Delta x}{2}\big)\Delta x$$

$$= U_n + \Big[a\big(U_n + f(U_n)\frac{\Delta x}{2}\big)^{\alpha} - b\big(U_n + f(U_n)\frac{\Delta x}{2}\big)^{\beta}\Big]\Delta x.$$

In Exercise 4, the reader is asked to carry out the midpoint method to this ODE.

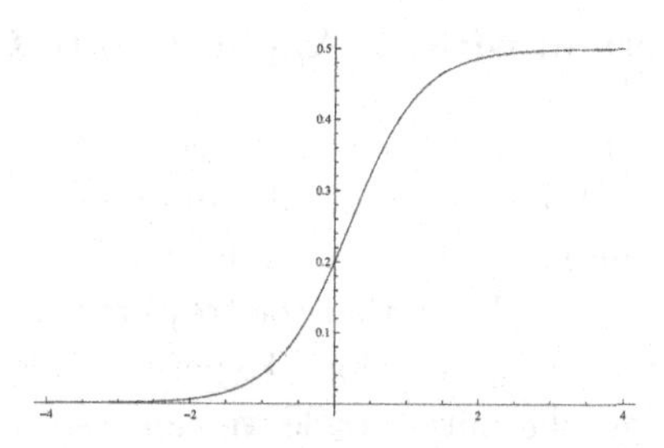

Figure 6.2.1: The sigmoidal curve, $U = 0.1e^{2t}/(0.5 + 0.2(e^{2t} - 1))$

In passing we remark that this model identifies a particular problem that arises in cancer therapy. Many drug therapies affect only the currently dividing subpopulation. Whereas, there is always a subpopulation behind this one. Once the surface cells have died, this reservoir of cells has access to the nutrient rich blood supply and begins to reproduce.

Exercises:

1. Continue the basic example from Section 6.1. By comparing the midpoint estimates for $u_3, ..., u_{10}$ against the actual data and the forward Euler estimates.

2. Apply the midpoint method to $U_t = 2U - U^2$, $U_0 = 100$ and $K = 2$. Compare your estimated results to the actual values for $U = 200e^{2t}/(2 + 100(e^{2t} - 1)$.

3. In (6.2.2) take $\alpha = 0.2$ and $\beta = 0.5$ with $K = 2$ and $U_0 = 0$. Use the midpoint method to estimate values of U. Plot the result. Is this a sigmoidal curve?

4. If the tumor is assumed spherical, then the researchers sometimes use the following versions of 6.2.2,

$$\frac{dU}{dx} = aU^{2/3} - bU.$$

Estimate values for U if $K = 0.5$ and $U_0 = 2$. Plot the points and use a B-spline to resolve the computed values as a curve.

5. Apply midpoint method to the birth/death process described in (6.1.2). Compare the midpoint method solution to the forward Euler solution.

6.3 Shooting Method with an Application to Cooling Fins

Many of us have had the experience of being on the athletic field and receiving a thrown, kicked or batted ball. In order to receive the ball, we must estimate the final location from the initial lift or angle, location and speed. From experience, we know that the trajectory is a parabolic arc, and of course we know the initial location. Experience also tells us the initial speed. When we know the initial angle we can resolve the trajectory and identify the final location. We also know that it takes some experience to do this reliably and quickly. This is the *shooting method.* We have been applying the technique since childhood. When the baseball announcer says that an outfielder has "a good jump on the baseball", he is saying that player solves the ODE faster than most.

We start with a second order ODE. The standard approach to a second order ODE is to reduce it with two first order equations which may be solved simultaneously. For instance given $u'' = f$, we set $v = u'$ and write the two equations, $v' = f$ and $u' = v$. To solve such a system, we would expect to be given two initial values, $u(0)$ and $v(0)$. We recognize this as an initial value problem. Another possibility is to be given values of u at the endpoints of the domain. This is called a *boundary value problem*. This is the problem that the baseball outfielder solves every day. In this section we present the shooting method for solving boundary value problems.

We begin with an example and develop the procedure within this context. Cooling fins are a biological adaptation that is copied in mechanical engineering designs. We are all used to the use of cooling fins on the cylinder head of internal combustion engines, for instance the cooling fins used on lawn mower or motorcycle engines. The idea is that the relatively large surface area compared to volume accelerates heat transfer from the object through the fin and then to the surrounding medium. Termite mounds in North Australia provide an example from nature. If you are not familiar with these structures, then please do a web search. In particular, these insect mounds use cooling fins to dissipate the heat from the colony to the surrounding environment.

We consider the following example based in a one dimensional model. Let $u = u(x)$ denote the temperature measured in absolute degrees or Kelvin at location x. Then u satisfies the following second order non-linear ODE.

$$u''(x) = k(u^4 - T^4), \tag{6.3.1}$$

where T is the temperature of the surrounding medium, L is the length of the fin, k is a heat flux constant and $x \in [0, L]$. The first task toward solving (6.3.1) is to rewrite it as a system of two first order equations,

$$u'(x) = v(x), \quad v'(x) = k(u^4 - T^4). \tag{6.3.2}$$

This now can be expressed as $u'(x) = f(x, v, u)$, $v'(x) = g(x, v, u)$, where

$$f(x, v, u) = v(x), \quad g(x, v, u) = k(u^4 - T^4).$$

Before continuing with the solution process, we consider the data. First if T were in normal range, say 27 C, then $T \approx 300$ Kelvin. If for instance, $u(0) = 35C = 308K$, then $v'(0) = 899,176,496$. Now, think about what will happen if we were to develop this pair of equations as an initial value problem using a simple procedure such as Euler forward. In this case, we would have a value for $v(0) = \alpha$ and write

$$v(0 + \Delta x) = v(0) + \Delta x * v'(0) = \alpha + 899,176,496(\Delta x),$$

$$u(0 + \Delta x) = u(0) + \Delta x * u'(0) = u(0) + v(0)\Delta x = 308 + \alpha\Delta x.$$

Now a serious problem occurs at the next iteration. Here we have

$$u(0 + 2\Delta x) = 308 + \Delta x(\alpha) + \Delta x(v(0 + \Delta x))$$

$$= 308 + \alpha\Delta x + \Delta x(\alpha + 899,176,496(\Delta x)).$$

In order to keep the values of u within reasonable limits, either α is a very large negative number, or Δx is very close to zero. The first case is inconsistent with the model. We expect u to be decreasing along the interval $[0, L]$. But a huge drop off in temperature is not intuitive. Alternatively, we must take Δx very close to zero. In terms of the given data, it must be a small multiple of 10^{-5}. Now, if $L = 0.5$, then we will need 50,000 iterations. For the purpose of this introductory presentation, this is far too many iterations. Hence, the data used below is intentionally not in normal range. When looking at the given data, keep in mind that even liquid nitrogen is 77 K.

We set $T = 3.0$, $k = 0.23$, $L = 0.25$ and $u(0) = 4.5$. As this is to be a boundary value problem, we must set $u(L)$. For instance, $u(L) = 3.0$. The shooting method proceeds as follows. First estimate $v(0)$ with $\bar{\alpha}$ and then use one of the standard techniques to derive $\bar{u}$ the computed estimate for u at L. If the estimate is not equal to 3.0, then we re-estimate with $v(0)$ with $\hat{\alpha}$ and calculate a new estimate $\hat{u}$ for u at L. The immediate goal is to get two estimates which bracket the desired value. With these two estimates for $v(0)$, we are now able to derive a third estimate using linear interpolation,

$$\alpha = \hat{\alpha} + (4.0 - \hat{u})\frac{\bar{\alpha} - \hat{\alpha}}{\bar{u} - \hat{u}}. \tag{6.3.3}$$

If the result is still not sufficiently close to the desired value, we repeat the process using the two prior estimates that best approximate for $u(L)$.

Historically, the shooting method was developed to aim a canon. The basic idea is that given the location of the cannon and the target, then the problem of aiming the cannon reduces to determining the cannon angle, $v = u'$. After two reasoned misses one far and one near, you are able to determine the correct canon angle, u' by interpolating between the two prior attempts.

We need to select a technique to compute the estimates. We choose the midpoint method for v and forward Euler for u. For these procedures, we have

$$v_{n+1} = v_n + h(x_n, v_n)\Delta x = v_n + g(x_n + \frac{\Delta x}{2}, v_n + g(x_n, v_n)\frac{\Delta x}{2})\Delta x$$

$$= v_n + k\big[u(x_n + \frac{\Delta x}{2})^4 - T^4\big]\Delta x = v_n + k\big[(u_n + v_n\frac{\Delta x}{2})^4 - T^4\big]\Delta x,$$

$$u_{n+1} = u_n + f(x_n, u_n)\Delta x = u_n + v_n\Delta x,$$

where v_n is the computed approximation of $u'(x_n)$. Notice that we used forward Euler to estimate a value of u when computing v_{n+1}.

Setting $\Delta x = 0.01$ and testing the result for negative integer values of α we see quickly that estimates of -10.0 and -11.0 bracket the desired result. The corresponding values for $u(L)$ are 3.18936 and 2.82282. Employing the interpolation procedure (6.3.3) we calculate the next test at -10.5166. This yields $u(L) \approx 3.00832$. The next application of (6.3.3) yields -10.5121 and $u(L) \approx 2.99996$ correct to 4 decimal points. Figure 6.3.1 and 6.3.2 show values for u and u' when $u'(0) = -10.5121$. Figure 6.3.2 shows $v' \to 0$ or the temperature decline is constant at the end of the fin.

If we had used midpoint method to estimate values of u, we would have calculated

$$u_{n+1} = u_n + h(x_n, u_n) = u_n + \big[f(x_n + \frac{\Delta x}{2}, u_n + f(x_n, u_n)\frac{\Delta x}{2})\big]\Delta x$$

$$= u_n + v(x_n + \frac{\Delta x}{2})\Delta x = u_n + \big[v_n + g(x_n, v_n)\frac{\Delta x}{2}\big]\Delta x,$$

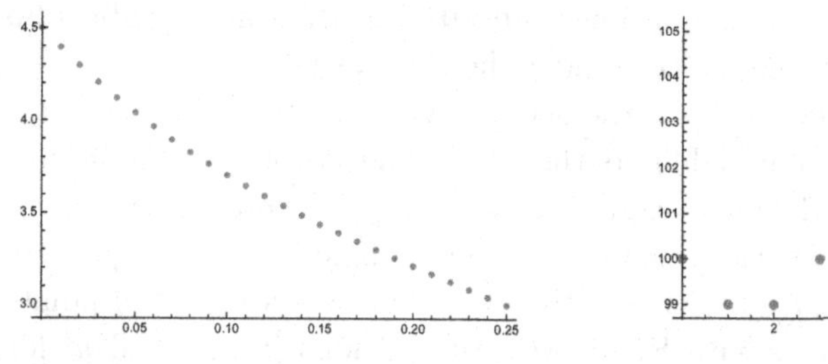

Figure 6.3.1: Temperature plot for shooting method

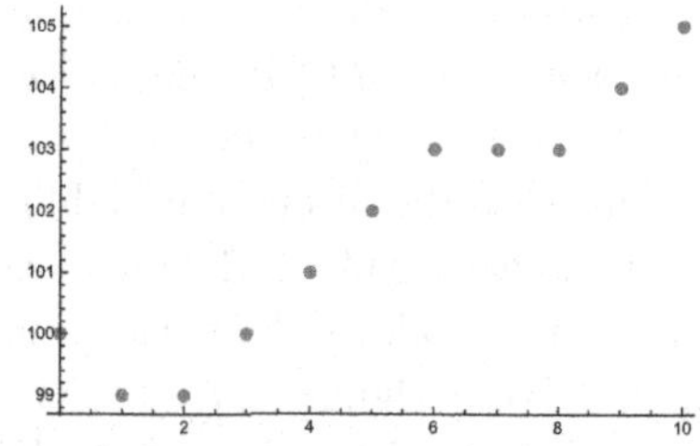

Figure 6.3.2: u' is asymptotic as $x \to 2.5$.

using forward Euler to estimate the needed value of v.

Exercises:

1. Execute the example from the text.

2. Repeat the example from the section using forward Euler to estimate u and v.

3. Execute the problem begun in 2. Set $L = 0.5$ and $u(79)$.

4. Use the shooting method to estimate the solution to the 1-D Helmholtz equation $u'' + u = 0$ on the interval $[0, 2\pi]$ with u(0) $= 0$ and $u(2\pi) = 0$.

5. The actual solution to 4 is $u(x) = \sin(x)$. Using your estimated solution compute the mean relative absolute error.

6. If you were to set up a problem using a cooling fin immersed in liquid nitrogen with $u(0) = 80$, what would be the appropriate value for Δx?

6.4 The Method of Lines, Revisiting the Heat Equation

There are two ways to look at *method of lines.* In either case you begin with a PDE for a transient, time dependent, process. One way to describe it is that you use some method to render the spatial part of the PDE. For instance, you apply FDM to the spatial part as in Section 4.2. In the literature, this form of the PDE is called the *semi-discrete form.* Then you recognize that the resulting equation is an ODE in the time variable. So you introduce an ODE method to render the time variable.

The other way to describe it is the one that gave rise to the terminology. In this case the discussion highlights the role of lines or planes. To provide context, we recall the 1-D heat equation, $u_t = \alpha u_{xx}$. The method of lines proceeds as follows. First the domain of u is a rectangle $D = [0, T] \times [0, 10.0]$ in $\mathbb{R}^2$. We designate Δt and Δx and thereby determine a lattice of points (t_n, x_i) in D. Exactly as with FDM, we write u_i^n for $u(t_n, x_i)$ and u^t for the state vector at time t. We will suppose that the x-axis partition has m subintervals.

Using the second central difference on the right hand side, we derive

$$\frac{\partial u_i(t)}{\partial t} = \frac{\alpha}{\Delta x^2}\left[u(t, x_{i+1}) - 2u(t, x_i) + u(t, x_{i-1})\right]$$

$$= \mu\left[u_{i+1}(t) - 2u_i(t) + u_{i-1}(t)\right], \tag{6.4.1}$$

where $\mu = \alpha/\Delta x^2$. On the one hand, we want to look at (6.4.1) as a statement about states. In particular, u would be a function of time taking values as $m+1$-tuples and (6.4.1) is a statement about the derivative of $u = (u_0, u_1, ..., u_m)$. Alternatively, we can look at it in terms of the coordinate functions, u_i for each $i = 0, 1, ..., m$. From this point of view, (6.4.1) is $m + 1$ ODE. The term, method of lines, arises in this last context. Indeed, each (t, x_i) describes a vertical line in the tx-plane.

Hence, we have arrived at an ODE where the left hand side is the derivative of u_i and the right hand side is a function of u_i, u_{i-1} and u_{i+1}. From here we may employ any of our ODE techniques. Should we choose to use forward Euler we have

$$u_i^{n+1} = u_i^n + \frac{\alpha \Delta t}{\Delta x^2}\left[u_{i+1}^n - 2u_i^n + u_{i-1}^n\right].$$

This we immediately recognize as FTCS FDM. Indeed, we see the source of the term, forward Euler, as applied to FDM. It comes from ODE methods via the method of lines.

Alternatively, consider the estimator/corrector. In this case

$$u_i^{n+1} = u_i^n + \frac{\alpha \Delta t}{2\Delta x^2}\left[(u_{i+1}^n - 2u_i^n + u_{i-1}^n) + (u_{i+1}^{n+1} - 2u_i^{n+1} + u_{i-1}^{n+1})\right].$$

We recognize this as Crank Nicolson. Hence, ODE time stepping overlaps with FDM time stepping. Midpoint method is yet another option.

If we had chosen to develop the method of lines with state functions, then we would arrive at the matrix form family from FDM.

The method of lines is very common. Often the spatial treatment is not FDM but finite element method, boundary element method or collocation method. No matter which spatial technique is used, time stepping via FDM or ODE methods is the standard.

Exercises:

1. Implement the method of lines for the one dimensional heat equation. Set $\alpha = 1/2$, boundary values $u(t,0) = 0$, $u(t,1) = 40$ and initial condition $u(0,x) = 0$ for $x < 1$ and $u(0,1) = 20$.

2. Implement the corrector for the setting given in Exercise 1.

3. Develop the 1-D heat equation with the method of lines with mid-point method of the ODE side.

4. Develop the first order wave equation $u_t = \alpha u_x$ with the method of lines. Use the setting given in Exercise 5 of Section 4.2.

   a. Use forward Euler.

   b. Use forward Euler with corrector.

Chapter 7

Monte Carlo Method

Introduction

In this chapter, we present an introduction to Monte Carlo method from the point of view of the numerical analyst. Monte Carlo method is a huge topic. We will only provide a small overview with emphasis on the numerical aspects of the topic. It would take a volume or more to present the technique even for a single application area.

A word of caution, the material of this chapter requires knowledge of probability distributions.

Monte Carlo method is a technique often used to solve problems too large or too complicated to be resolved by direct means. In this regard, it is often referred to as the method of last resort.

Consider an example. Suppose you want to drive across town. There are several alternate routes and combinations of alternatives. For some, traffic flow is faster but the route is longer, for others, some intersections may be clogged due to heavy traffic density and narrow streets. Of course, all of this depends on the time of day and even the time of year. Some of the events are dependent on others and some are probabilistically independent. How do you determine on a given day at a given time what is the optimal route? We might estimate the time required to complete each individual segment of the journey as a probability distribution. But it is not reasonable to suppose that each segment will perform at the distribution mean when combined with the others. There may be dependencies between the distributions. Hence, it is not possible to do the problem via a direct calculation. It is equally impossible to send ten thousand cars through the system and then calculate the mean travel time for each possible alternative. But we can simulate the system on a computer. When we do that ten thousand, a hundred thousand, a million times and then calculate the mean travel time for each alternative, we are using *Monte Carlo method.*

There are aspects to the example that are specific to the example and aspects that are generally associated to Monte Carlo. In addition, the technique may vary from one application area to the next. Nevertheless, there is a commonality that determines Monte Carlo.

The basic idea is to represent a process or system on a computer. The representation should be as correct as possible. Some aspects may not be known to us in a deterministic manner. They may be simple but unknown or too complex to be known precisely. In these cases the state of some system components will need to be estimated via a probability distribution. When this is done, you can establish instances for each system segment. Next, you execute a particular combined system instance. Having done it once, you repeat the process over and over again. Each time you use a randomly derived instance for each unknown segment. Finally, you can reliably estimate the target value. Indeed, you will know the system by observing how it functions at the macro level rather than by determining the exact interaction of each constituent part at the micro level. It is like treating the system like a black box.

The difficulty in implementing the Monte Carlo method lies in first estimating the probability distributions for the constituent parts. And when that is done, then to determine a random state of the system. And of course we must do this thousands of times. Hence, we must have efficient means to generate random numbers that are distributed in any number of ways. We will see that this is by no means easy. Even with *Mathematica* to help us over large parts of the problem, there are still cases that we must compute.

So far we have only mentioned problems where certain segments of the problem are unknowable or too complex to be known specifically. There is another setting that is often resolved with Monte Carlo method. Suppose we have a PDE that we expect to use Crank Nicolson FDM to estimate values of the solution function. However, it may be that the x-axis partition is so small that the matrices are too large to invert. Even advanced large matrix processes will not resolve the inverse. For this setting, we have no choice but to use Monte Carlo.

Monte Carlo method began during WWII with an impossibly complex question from quantum mechanics. Stanislaw Ulam suggested doing repeated simulations. Jon von Neumann recognized the generality of the process. The two very quickly set up a project. As with any project during the war, it needed a code name. It was titled *Monte Carlo.* Hence, the name we use today.

This chapter develops the rudiments of Monte Carlo in 4 sections. We need to review some of the terminology from probability theory. We do this in the first section. We assume that the reader is familiar with probability theory. In the first section, we will set the scene for Section 2, the development of efficient pseudo-random number generators. In Section 3, we consider Monte Carlo as applied to some simple problems such as integral (area) estimation. Finally, in Section 4 we look at the problem of estimating the solution to a PDE. There are two types of PDE of particular interest, PDE where some of the parameters may be known only as random variables and PDE where the dependent variable is random. We consider both in this section.

One final remark, Monte Carlo method is rarely efficient. If there is an alternative, it is likely more efficient and hence, preferred.

7.1 Basics of Probability Theory, Terminology and Notation

In this section we present some of the basic terminology and notation for probability theory. Our primary purpose is to standardize. We suppose throughout that the reader is already familiar with the concepts found here. To start, we do not want to develop measure theory and sigma fields. Hence, we will be more intuitive than precise in the beginning. Furthermore, we want to emphasize continuous variables. We include discrete variable concepts as needed in subsequent sections. For a more detailed and more general development see [Olkin, Gleser and Derman (1980)].

We take Σ as a set of events and a real valued $X : \Sigma \to \mathbb{R}$. The function X is called a *random variable* provided there is a *probability density function* f_X. In particular, the function f_X is real valued and defined in a possibly infinite interval (a, b) that contains the image of X. Further, f_X is

- continuous,
- non-negative,
- with image contained in (a, b),
- integrable, $\int_a^b f_X(x)dx < \infty$.

Without loss of generality, we can normalize so that the value of the integral is 1. Next, we say that the probability of an event ξ satisfying

$X(\xi) \in (c, d) \subset (a, b)$ is given by

$$P(c < X < d) = \int_c^d f_X dx. \tag{7.1.1}$$

It may be helpful to think of $P(a < X < d)$ as the measurement of the area of $X^{-1}(c, d) \subset \Sigma$ relative to the area of Σ.

Continuing, we use the density function to define the *distribution function*. In particular, $F_X(x) = P(X < x)$. It is immediate that

$$F_X(x) = \int_a^x f_X.$$

As f_X is continuous, then F_X is differentiable (the fundamental theorem of calculus) with derivative f_X. Since f_X is non-negative, then F_X is increasing and hence, injective.

Recall that sometimes the distribution function is called the *cumulative distribution function.* Furthermore, it is standard to use the term *distribution* to refer to the setting in its entirety. In particular, we will refer to X as a distribution. This is a short form of the statement that X is a random variable with density and distribution functions.

The two most important quantities associated to a distribution are the *mean* $E[X]$ or μ_X and the *variance* $var[X]$ or σ_X^2. These values are given as follows,

$$E[X] = \int_a^b x f_X dx \tag{7.1.2}$$

and

$$var[X] = \int_a^b (x - X[x])^2 f_X dx. \tag{7.1.3}$$

The square root of the variance is called the *standard deviation.* Notice that the variance may also be defined as $E[(X - E[X])^2]$. Also sometimes the mean is referred to as the *expected value.*

Continuing, given two distributions associated to random variables X and Y, the *covariance* is given by

$$\sigma_{XY} = E[(X - E[X])(Y - E[Y])].$$

Hence, the covariance of X and X is the variance of X. The covariance leads to the *correlation* defined by

$$\rho_{XY} = \sigma_{XY}/\sigma_X\sigma_Y.$$

Finally, we say that the distributions X and Y are *independent* provided $\rho_{XY} = 0$.

The particular case of $f_X(x) = 1/(b-a)$ for a finite interval (a, b) is called the *uniform distribution*. It is immediate that for the uniform distribution satisfies

$$P(c < X < d) = \int_c^d f_X dx = \frac{d-c}{b-a}.$$

In other terms, the probability that X lies in any interval depends only on the size of the interval. Next, we say that a sequence of numbers $\{x_n\} \subset (a, b)$ is *random* provided that for each n, $P(c < x_n < d) = (d-c)/(b-a)$. The uniform distribution gives rise to the idea of a random number sequence.

To be precise, we say that the sequence $\{x_n\}$ is *random* if for each n, x_n is uniformly distributed and any pair x_n and x_m are independently distributed. In the next section we discuss the problem of generating random number sequences.

Exercises:

1. Prove that the random variables that the set of random variables defined on Σ forms a vector space.

2. Given two distributions X and Y, prove that $X = \alpha Y + \beta$ if and only if $\rho_{XY} = \pm 1$.

3. Prove that the expected value, E is a linear function.

7.2 Generating Distributed Number Sequences

In order to use the Monte Carlo method, it is necessary to have sequences of random numbers. Technically stated, they must be sequences x_n with the property that the individual entries are uniformly distributed and any two entries are independently distributed.

This has been a problem since people started using the method. Initially, they tried to generate random number sequences by observing natural phenomenon. For instance, they observed the arrival of particles from the sun or other extraterrestrial source and counted arrivals to a given location during a prescribed time interval. There were two problems with these procedures. First, they could not prove that the sequences were in fact uniformly and independently distributed. Second, no sequence could be repeated. So they could never rerun a simulation. These problems lead to the development of *pseudo-random numbers*.

The traditional approach to generate a pseudo-random number sequence uses a simple result of elementary group theory. We start with a prime number m, an integer a not congruent to 0 or 1 $\mod m$ and a seed integer x_0 also not congruent to 0 or 1 $\mod m$. We proceed recursively, setting $x_1 = a*x_0 \mod m, ..., x_{n+1} = a*x_n \mod m$. By elementary group theory, this results in a listing of the numbers between 1 and $(m-1)$ in some order depending on a. If m is relatively large compared to n, then the sequence will appear to be uniformly and independently distributed. But of course, once x_0 is chosen, the entire sequence is predetermined. This is the source of the term *pseudo-random* because they appear random but they are in fact generated using a completely deterministic procedure.

This technique for generating pseudo-random number sequences is called the *linear congruence method.* To be useful, m needs to be a very large prime number. For instance, $m = 2^{18837} - 1$. To make it a little more difficult to identify a and m, they often use an integer c to offset the sequence as follows.

$$x_{n+1} = a * x_n + c.$$

The source of the numbers can be further camouflaged by reporting values modulo some $k < m$.

Up to the mid 1990s, *Mathematica* and competing 4GL math/stat systems used linear congruence as the primary means for generating pseudo-random numbers. To make output appear random they would use some combination of the date/time for the seed. Of course, if you knew the seed, then the sequence was reproducible. This was desirable as it allowed for system testing.

There was a problem. The sequence would always repeat and if you knew the first repeating value, then it was a simple matter to determine the seed as well as a and m. Thus, they had limited use as a security code generator. For that and other reasons, linear congruence is no longer widely used.

One of the procedures in current use is called the *Mersenne Twister.* The base idea is to do linear congruence on vectors. In particular, we generate n pseudo-random numbers via linear congruence to form an n-tuple, v. Next, we take an $n \times n$ nonsingular matrix T and report the sequence y_i as the ith entry of $w = Tv$. With an offset vector this becomes,

$$w = Tv + \gamma.$$

A common choice for T is defined as follows.

$$y_0 = x_1, ...y_i = x_{i+1}, ...y_n = \sum_{i=0}^{n} a_i x_i,$$

for preselected integers a_i. Note that this transformation arises naturally when developing canonical forms.

To make the underlying process more opaque, you can permute the entries of Tv with a permutation matrix A. Recall that a permutation matrix is one with exactly one nonzero entry in each row and column. The nonzero entry is a 1. Multiplication by A will permute the entries of Tv. This is called the *Mersene twister with shift register.* One problem with linear congruence was that no integer appears more than once except when the entire sequence repeats. The Mersene twister allows for repetitions in a manner that is difficult to predict.

All of these methods are available to the *Mathematica* user. In addition, you can specify any of the parameters including the seed. This is well described in the system documentation.

There are yet two more alternatives available in *Mathematica.* If you use an Intel processor, there is a hardwired pseudo random number generated called MKL. You access this feature in *Mathematica* as *IntelMKL.*

Finally, if you replace the linear transformation T in the Mersene twister with a more general function, then the procedure is called *extended CA* or extended cellular automation. Variations of Extended CA are available via *Mathematica* for specified distributions.

We have yet to discuss generating sequences with a particular nonuniform distribution. The common procedure is called the *inverse distribution method.* The procedure is conceptually simple and available in *Mathematica.*

We begin with a variable X with cumulative distribution function F_X. Recall that the definition of the distribution function, $F_X(\alpha) = P(X < \alpha) = x$. Hence, F_X is increasing and it follows that F_X is injective. In particular, for any x in $[0, 1]$, the image of F_X, there is a unique α preimage of x.

We claim that if $\{x_n\}$ is a pseudo-random uniformly distributed sequence, then $\{\alpha_n\}$ is a random sequence distributed as X. Indeed, suppose the U is the uniform variable, $U = F_X \circ X$, then $F_X^{-1} \circ U$ is X. The claim follows.

We mentioned that *Mathematica* provides random sequences distributed in the commonly occurring distributions. This is especially important as we cannot usually derive the inverse distribution function. When we must have values for the inverse of F_X, we compute *quantiles* and use those to interpolate values as needed.

Quantiles are determined as follows. Given a random variable X, the α percent quantile is x so that $P(X < x) = \alpha$. For instance, the distribution *median* is the 50 percent quantile, the value of x so that $P(X < x) = 0.5$. Quantiles are usually estimated for an unknown distribution form sampling data.

Finally, given a set of quantiles, you estimate the distribution using the B-spline fit. With this representation it is routine to compute preimages for selected x.

Exercises:

1. Suppose that X is the distribution with density function f, where

$$f(x) = 0, \ \ x < -1, \ \ f(x) = x - 1, \ \ x \in [-1, 0],$$

$$f(x) = 1 - x, \ \ x \in [0, 1], \ \ f(x) = 0, \ \ x > 1.$$

a. Derive the cumulative distribution function F_X.
b. Derive F_X^{-1}.
c. Compute a random sequence of 500 reals with distribution X.

2. Repeat Exercise 1 but do not derive the distribution function, rather, estimate quantiles at $0.1, 0.2, ..., 0.9$ and use cubic B-splines to fit the data. What is the error?

3. Use *Mathematica* to generate sequences of random numbers that are uniformly, normally (mean 0, variance 1) and exponentially (mean 1) distributed. Plot each sequence. Set the sequence lengths to be 500.

4. The log normal distribution is important in finance. If X is a normal variable, then e^X is log normal. Plot a sequence of randomly generated log normal numbers. Set the sequence length at 100.

7.3 Monte Carlo Method for Estimating the Integrals and n-Dimensional Volumes

The first thing to be noted is that we rarely need Monte Carlo techniques to estimate integrals of functions mapping $\mathbb{R}$ to $\mathbb{R}$. The techniques in Chapter 5 are sufficient. Nor do we need these procedures for functions defined on n-dimensional cubes in $\mathbb{R}^n$, the techniques from Chapter 5 are equally adequate for these cases.

As integrals and volumes are closely related, we choose to concentrate on integrals. We include an area problem in the exercises.

Keep in mind that Monte Carlo method for estimating integrals applies only to cases where we can evaluate the function at nearly every point in the domain.

The basic idea for Monte Carlo integration rests on the mean value μ of f. Suppose we want to integrate f over a region A of volume $V(A) = \int_A 1dV$, then μ is given by

$$\mu = V(A) \int_A f dV.$$

In turn, if $x_1, ..., x_N$ is a random number sequence, then

$$\mu \approx \frac{1}{N} \sum_{i=1}^{N} f(x_i).$$

Therefore,

$$\frac{V(A)}{N} \sum_{i=1}^{N} f(x_i) \tag{7.3.1}$$

estimates the integral. We refer to the method given in (7.3.1) as basic Monte Carlo integration.

The procedure is convergent in most cases. Indeed, the error depends on the variance of f. If we estimate the variance with the sample variance σ_N^2, then the error is the limit of $V(A)\sigma_N/\sqrt{N}$ as N goes to ∞. Hence, the error converges to zero provided $\{\sigma_N\}$ is bounded. There is a second consequence of this calculation. We halve the error by quadrupling the size of the sample. This is further evidence that Monte Carlo is not efficient. It is not the technique of choice. Rather, we use Monte Carlo when there is no viable alternative.

There are things that can be done to improve the efficiency of the Monte Carlo integration. For instance, consider the following procedure, referred to as *Monte Carlo integration with stratified sampling.* The goal here is to divide the domain into subregions so that the function variance is smaller on each subregion. Then, since convergence depends on the variance, we expect faster convergence when estimating the integral on the subregions.

Suppose we are iterating an integral estimation procedure while keeping track of the estimated error. We plan to continue the procedure as long as the error estimate is greater than a preset kickout threshold. As we proceed, it may be that the rate of convergence seems to be slow. In this event, we consider subdividing the function domain.

We can divide the region along coordinate axes, planes, hyper-surfaces and test the variance of f on each pair of subdomains. For instance, in $\mathbb{R}^n$ there are n possible pairs of subdomains each corresponding to a coordinate hypersurface. In each case, we estimate the local variance of f using the fraction of the current random sample that lies in that subdomain and the usual sample variance. Next, we select the pair for which the sum of the two sample variances is minimal and use Monte Carlo integration (7.3.1) individually on each. We can repeat the domain partitioning process as needed to increase the rate of convergence. If we repeat k times, then the final result will be the sum of 2^k subintegrals.

In this manner, we get maximal effect from our repeated calculations. Of course, it is important to keep in mind that 2^k grows very fast. Any automated application must limit the number of iterations of the partitioning process.

Another variation of the basic Monte Carlo integration is based on a significance criteria. Given a function f defined on a domain D, we seek a decomposition $D = E \cup F$, $E \cap F = \emptyset$. We choose E and F so that $|f|$ is relatively large on E and small F. For instance, suppose f is the density function for a multivariate normal distribution of mean 0 and variance 1. Most of the integral of f over $\mathbb{R}^n$ is within the n-sphere of radius 2. Now, take E to be this sphere and $F = \mathbb{R}^n - E$. Hence, we should over sample on E and under sample on F.

Exercises:

1. Estimate the integral of $e^{|x+y|^2}$ over the unit circle.

2. Use the integral estimation technique to estimate the volume of a sphere in 3 dimensions. Recall that a sphere in $\mathbb{R}^3$ is a set of point (x, y, z) with $x^2 + y^2 + z^2 \leq 1$.

a. Estimate the area by estimating the integral of $f(x, y, x) = [x^2 + y^2 + z^2]^{1/2}$ with Monte Carlo method as presented in the section.

b. Estimate the area but taking a random sample (x_n, y_n, z_n) with $-1 < x_n, y_n, z_n < 1$. This latter condition determines a cube of volume 8 around the sphere. Next, if s is the number of sample points inside the sphere and t is the size of the sample, then $8s/t$ is an estimate for the volume of the sphere.

3. Repeat Exercise 2 for a sphere in 4 dimensions.

7.4 Monte Carlo Method for Problems with Stochastic Elements

In this section, we consider three types of problem. We develop each with an example.

We begin with a telephone call system. This sort of problem is related to the *cross town trip* problem discussed in the chapter introduction.

Suppose we have 5 operators to answer calls, that a call lasts 3 minutes and arrives on average every 30 seconds (0.5 minutes). When a call arrives, then it will be assigned to an available operator. Otherwise, the call will be held until an operator comes free. On the other hand, if there are more operators than current calls, some operators will remain idle. The goal is to design the system to minimize two competing goals, the average wait time and the daily operator idle time.

For this setting, it is usual to take the inter-arrival time to be *exponentially distributed*. In this case $f_X(x) = \lambda e^{-\lambda x}$ and $F_X(x) = 1 - e^{-\lambda x}$ where λ is the expected value. It is now a simple matter to generate a pseudo-random sequence $\{x_i\}$ of reals with $\lambda = 0.5$. If we set $y_n = \sum_{i=1}^{n} x_i$, then we have a random sequence of arrival times. Alternatively, *Mathematica* provides a command that will return an exponentially distributed sequence.

To simulate the system, we need to set the time step size, $\Delta t = 0.1$, a list $p_1, p_2, ..., p_5$ to represent the operators, and two variables w and s, one to represent the size of the wait queue and one to count the number of idle operators. The variable p_i is set to 3 when the operator is assigned a call and decremented at each time step. When $p_i = 0$, we know that the operator is available to take a call.

At each time step, we must maintain each variable. For instance, at the mth time step we need to

- determine k the number of calls that arrived between $(m-1)\Delta t$ and $m\Delta t$,

- set $w = k + w$,
- update each operator, compute $p_i = p_i - \Delta t$ for each positive p_i,
- if $w > 0$ and $p_i = 0$, set $p_i = 3$ and subtract 1 from w. Repeat this step until $w = 0$ or all $p_i > 0$.
- do summary statistics, accumulate $w * \Delta t$ to the total caller wait time and $s * \Delta t$ to operator down time.

The goal is to balance the total caller wait time and the total operator down time. For instance, the result of the study may be to alter the number of operators to improve the efficiency of the system. In particular, we may set a maximal customer wait time goal and then determine the minimal number of operators necessary to achieve that goal.

Next, we look at a PDE with a random parameter. Consider the Black Scholes equation

$$\frac{\partial V}{\partial t} + \frac{1}{2}\sigma^2 S^2 \frac{\partial^2 V}{\partial S^2} + rS\frac{\partial V}{\partial S} - rV = 0. \tag{7.4.1}$$

For a derivation of this equation see [Loustau (2016)]. The equation relates the price of a stock option, V as a function of time t and the spot price of the underlying security S. The parameter r is the base interest rate during the life of the option and σ measures the volatility of the security price.

Usually, r is pegged to the rate of the 30 year US Treasury bill. The board of the Federal Reserve meets monthly and at each meeting determines the base interest rate for the US banking system. Hence, it is reasonable to reject the assumption that r is constant for the life of a 6 month contract. Since we do not expect economic conditions to dictate both up and down movements in r, then we would choose a distribution X with non-negative f_X. For instance, we could randomly select historical six month intervals and use that data to infer a distribution for r. Hence, we denote it as r_t.

Once we have selected a distribution for r_t, we estimate the distribution inverse function and use that to generate a sequence $\{r_n\}$ of interest rates at each time period. If we chose to implement FDM with backward time and central space, then (7.4.1) with standard FDM notation becomes

$$\frac{V_i^n - V_i^{n-1}}{\Delta t} + \frac{\sigma^2}{2} S_i^2 \frac{V_{i+1}^n - 2V_i^n + V_{i-1}^n}{\Delta S^2} + r_n S_i \frac{V_{i+1}^n - V_{i-1}^n}{\Delta S} - r_n V_i^n = 0.$$

Setting $\lambda = \sigma^2 \Delta t / 2\Delta S^2$ and $\mu = \Delta t/\Delta S$, this equation yields the following implicit formulation

$$V_i^{n-1} = (\lambda S_i^2 + \mu r_n S_i)V_{i+1}^n - (1 + 2\lambda S_i^2 + r_n)V_i^n + (\lambda S_i^2 - \mu r_n S_i)V_{i-1}^n.$$

It is important to notice that the time stepping matrix changes with the time step. This is different from the FDM implementations we saw in Chapter 4.

We solve the equation as in initial value problem starting at the end state, $t = 0.5$ and stepping back to $t = 0$. Indeed, the value of the option is known at the end state, it is equal to the S provided $S > K$, and it is equal to zero if $S \leq K$. Whereas, we want to know V^0, the value of the option now. We repeat this process for different interest rate configurations. The Monte Carlo solution is the mean value of V^0 for repeated interest rate instances.

An alternative approach for problems of this sort is stochastic collocation. In many settings that procedure will be more efficient. See [Xiu (2010)].

Next, we consider a random variable X_t with mean μ and variance σ^2 and a diffusion process

$$dX_t = \mu(X_t)dt + \sigma^2(X_t)dW_t \tag{7.4.2}$$

where W_t is a Wiener process. [McLeish (2005)] In particular, for any Weiner process, $(W_{t+u} - W_t)$ is normally distributed with mean zero and variance depending on u.

We use the forward Euler method to develop a Monte Carlo simulation of the process in 7.4.2. We set a value for Δt and a randomly generated sequence $\{w_n\}$ of normally distributed values with mean 0 and variance Δt. For initial time t_0, we write

$$X_{t_0+n\Delta t} = X_{t_0+(n-1)\Delta t} + \mu(X_t)\Delta t + \sigma(X_t)w_n.$$

Now, repeat the time stepping process for successive values of n. This process is analogous to doing forward Euler method for an ODE.

In the context of the material presented here, these three problems are conceptually routine. However, they are anything but simple.

Exercises:

1. Complete the call center example for the text.

2. Redo the call center problem with the following data change. Suppose that the call length is $3 + \delta$ where δ is a normally distributed random variable with zero mean and variance 0.01.

3. Set up the matrix for explicit FDM with the Black Scholes equation (7.4.1).

4. Execute the Monte Carlo process with forward Euler for the equation 7.4.2. Set $\Delta t = 0.001$, $\mu = \sigma = 1$.

Bibliography

Atkinson, K. E. (1989). *An Introduction to Numerical Analysis*, 2nd edn. (J. Wiley).

Golub, G. H. and Van Loan, C. F. (2012). *Matrix Computations*, 4th edn. (Johns Hopkins).

Grcar, J. P. (2011). John von Neumann's Analysis of Gaussian Elimination and the Origins of Modern Numerical Analysis, *SIAM Review* **53** pp 607-682.

Hildebrand, F. B. (1974). *Introduction to Numerical Analysis*, 2nd edn. (Dover).

Liesen, J. and Strakos, Z. (2015). *Krylov Methods: Principles and Analysis* (Oxford Univ. Press).

Loustau, J. and Dillon M. (1993). *Linear Geometry with Computer Graphics* (Marcel Dekker).

Loustau, J. (2016). *Numercial Differential Equastions, Theory and Technique, ODE Methods, Finite Differences, Finite Elements and Collocation* (World Sceinctific Press).

Marsden, J. E. and Tromba, A. J. (2003). *Vector Calculus*, 5th. edn. (W. H. Freeman).

McLeish, D. L. (2005). *Monte Carlo Simulations and Finance* (J. Wiley).

Olkin, I,. Gleser L. J. and Derman, C. (1994). *Probability Models and Applications* (Prentice Hall).

Rudin, W. (1976). *Elements of Real Analysis*, 3rd. edn. (McGraw-Hill).

Schoenberg, I. J. (1973). *Cardinal Spline Interpolation* (SIAM).

Simmins, G. F. and Robertson, J. S. (1991). *Differential Equations with Applications and Historical Notes*, 2nd edn. (McGraw-Hill).

Su, B., Q. and Liu, D. Y. (1989). *Comptational Geometry: Curve and Surface Modeling* (Academic Press).

Thomas, J. W. (1999). *Numerical Partial Differential Equations, Conservation Laws and Elliptical Equations* (Springer).

Xiu, Dongbin (2010). *Numerical Methods for Stochastic Computations: A spectral Method Approach* (Princeton).

Wodarz, D. and Komarova, N. L. (2014). *Dynanmics of Cancer, Mathematical Foundations of Oncology* (World Scientific Press).

Index

Printed in the United States
By Bookmasters